Acknowledgement

This symposium was sponsored
by the Japan Society for the Promotion of Science
and this volume was edited for
the Restoration Medical Engineering Program.

Tissue Engineering for Therapeutic Use 3

Tissue Engineering for Therapeutic Use 3

Proceedings of the Third International Symposium of Tissue Engineering for Therapeutic Use, Tokyo, 4—5 September 1998

Editor:

Yoshito Ikada
Research Center for Biomedical Engineering
Kyoto University, Kyoto, Japan

Co-editor:

Teruo Okano
Institute of Biomedical Engineering
Tokyo Women's Medical University, Tokyo, Japan

1999

ELSEVIER

Amsterdam – Lausanne – New York – Oxford – Shannon – Singapore – Tokyo

ELSEVIER SCIENCE B.V.
Sara Burgerhartstraat 25
P.O. Box 211, 1000 AE Amsterdam, The Netherlands

First edition 1999

Library of Congress Cataloging in Publication Data
A catalog record from the Library of Congress has been applied for.

International Congress Series No. 1175
ISBN: 0-444-50029-4

⊗ The paper used in this publication meets the requirements of ANSI/NISO Z39.48-1992 (Permanence of Paper).

Printed in the Netherlands

Preface

Tissue engineering has the potential for having a dramatic effect on the practice of medicine by providing functional tissues and organs for tissue regeneration and replacement. However, successful growth of this interdisciplinary field requires a focused integration of engineering, life sciences, and clinical medicines to develop advancements in biomaterials, bioreactor developments, biomechanics, optimization of tissue integration and function, and predictive modeling for the design of future physiologic replacement parts.

The main purpose of our symposium is the same as that of the First and Second International Symposia of Tissue Engineering for Therapeutic Use, that is, to bring together biologists, material scientists, and clinicians with interests in tissue engineering from all over the world to review and discuss the latest developments and approaches related to tissue engineering. Another aim of this symposium is to inform the participants of the advances made in the Tissue Engineering Project of the Research for the Future Program sponsored by the Japan Society for the Promotion of Science (JSPS), which began in fiscal 1996.

Tissue engineering, which aims at regenerating new tissues as well as substituting lost organs by making use of autogeneic or heterogeneic cells in combination with biomaterials, is a newly emerging biomedical engineering field. There are several driving forces that presently make tissue engineering very challenging and important: 1) the limitation in biological functions of current artificial tissues and organs made from man-made materials alone, 2) the shortage of donor tissues and organs for organ transplantation, 3) recent remarkable advances in regeneration mechanisms made by molecular biologists, as well as 4) achievements in modern biotechnology for large-scale tissue culture and growth factor production. These current trends in life sciences and biotechnology will open up new possibilities for biomaterial research which has mostly until now focused on the biological performance of metals, ceramics, and synthetic polymers, as well as their interaction with components of the biological system such as proteins, cells, and tissues. However, a number of hurdles must be overcome before tissue engineering can actually make a substantial contribution to clinical medicine, without which all of these research endeavors will be in vain. A source of cells that are applicable to patients and can be preserved for long periods of time without severe cellular damage must be developed. In addition, we have to be able to synthesize artificial extracellular matrices to provide the cells with optimum environments for cell attachment, proliferation, and differentiation. Furthermore, controlled release of proper growth factors into the cells will be required for the promotion of tissue regeneration and augmentation.

Finally, I, as the Project Chairman, wish to express appreciation to coauthors, who have contributed their knowledge, experience, and time in the preparation of this volume.

June 9, 1998

Chairman of "Tissue Engineering Project"
Yoshito Ikada
Professor
Institute for Frontier Medical Sciences
Kyoto University

Tissue Engineering Project of the Research for the Future Program
Japan Society for Promotion of Science

Research project	Project leader
Biomaterials for tissue engineering	Teruo Okano (Tokyo Women's Medical College)
Bioprocess engineering of functional regeneration of cultured cells	Norio Oshima (University of Tsukuba)
Tissue engineering for soft tissues	Yasuhiko Shimizu (Kyoto University)
Tissue engineering for organs	Yoshio Yamaoka (Kyoto University)
Tissue engineering for hard tissues	Shoji Enomoto (Tokyo Medical and Dental University)
Functional regeneration by medical bioengineering	Tomomitsu Hotta (Tohkai University)

Contents

Preface — Y. Ikada   v

JSPS Tissue Engineering Project: Functional regeneration by medical bioengineering
Ex vivo expansion of human cord blood stem cells and its application to gene therapy
    K. Ando, H. Kawada, T. Shimizu, T. Tsuji, Y. Nakamura, M. Kimura,
    H. Miyatake, Y. Shimakura, S. Inokuchi, S. Kato and T. Hotta   1
Expansion of hematopoietic stem cells — new possibilities
    C. Eaves, E. Conneally, J. Audet, S. Rose-John, A. Eaves and J. Piret   15
Signal transduction of vascular endothelial growth factor (VEGF) receptors, Flt-1 and KDR/Flk-1
    M. Shibuya, T. Takahashi, A. Sawano, S. Hiratsuka, S. Ogawa, N. Yabana,
    Y. Maru, T. Noda and S. Yamaguchi   25
Liver stem-like (oval) cells: isolation, differentiation and application for cell therapy
    T. Sugiyama, O. Yasui and N. Miura   35
Development of a hybrid artificial liver support system and its application to hepatic failure animals
    K. Nakazawa, H. Ijima, M. Kaneko, J. Fukuda, T. Gion, M. Shimada,
    K. Shirabe, K. Takenaka, K. Sugimachi and K. Funatsu   43
JSPS Tissue Engineering Project: Bioprocess engineering of functional regeneration of cultured animal cells
Improvement of the performance of a packed-bed bioartificial liver
    N. Ohshima, K. Yanagi, H. Miyoshi and P. Kan   53
Repair of adult rat corticospinal tract by transplants of olfactory ensheathing cells
    G. Raisman   65
JSPS Tissue Engineering Project: Tissue engineering for soft tissues
Regeneration of the peripheral nervous system by artificial nerve conduit
    K. Suzuki, K. Ohnishi, T. Kiyotani, G. Lee, A.K. Kitahara, Y. Suzuki,
    K. Tomihata, M. Teramachi, Y. Takimoto, T. Nakamura, K. Endo,
    Y. Nishimura, Y. Shimizu and Y. Ikada   71
JSPS Tissue Engineering Project: Tissue engineering for organ regeneration
Retinal transplantation of clonal adult rat hippocampus-derived neural stem cells
    M. Takahashi, A. Nishida, I. Nakano, J. Takahashi, A. Mizoguchi, C. Ide
    and Y. Honda   77

Scaffold structure for a bioartificial liver support system
J. Mayer, E. Karamuk, K. Interewicz, T. Akaike and E. Wintermantel   87
JSPS Tissue Engineering Project: Biomaterials for tissue engineering
Novel manipulation technology of cell sheets for tissue engineering
M. Yamato, A. Kikuchi, S. Kohsaka, T. Terasaki, H.A. von Recum,
S.W. Kim, Y. Sakurai and T. Okano   99
JSPS Tissue Engineering Project: Biomaterials for tissue engineering
Reconstituted collagen assemblies as building blocks for the construction
of multicellular system in vitro
T. Hayashi, M. Hirose, M. Yamato, K. Mizuno, K. Nakazato, E. Adachi,
H. Kosugi, Y. Sumida, T. Okano, K. Yoshikawa, Y. Takeda, S. Takahashi
and Y. Imamura   109
Biocompatible alginate scaffolds enabling prolonged hepatocyte functions
in culture
R. Glicklis, S. Zmora, L. Shapiro, S. Cohen, R. Agbaria and J.C. Merchuk   119
Engineering of cell lines for diabetes therapy
C.B. Newgard, C. Quaade, A. Thigpen, H.E. Hohmeier, V. Vien Tran,
F. Kruse, H.-P. Han, G. Schuppin and S. Clark   133
Bone-inducing implants: new synthetic absorbable poly-D,L-lactic
acid-polyethylene glycol block copolymers as BMP-carriers
K. Takaoka, N. Saito, S. Miyamoto, H. Yoshikawa and T. Okada   141
JSPS Tissue Engineering Project: Tissue engineering for hard tissues
Tissue engineering in bone
M. Noda, K. Tsuji, T. Yamashita, N. Kawaguchi, Y. Ezura, J. Li,
S. Murakami, I. Sekiya, Y. Asou, Y. Takazawa, K. Furuya, Y. Liu
and A. Nifuji   153

Index of authors   159

Keyword index   161

1

JSPS TISSUE ENGINEERING PROJECT: FUNCTIONAL REGENERATION BY MEDICAL BIOENGINEERING

Ex vivo expansion of human cord blood stem cells and its application to gene therapy

Kiyoshi Ando[1,2], Hiroshi Kawada[1,2], Takashi Shimizu[1,3], Takashi Tsuji[5], Yoshihiko Nakamura[1], Minoru Kimura[4], Hiroko Miyatake[1], Yasuhito Shimakura[1,2], Sadaki Inokuchi[1], Shunichi Kato[1,3] and Tomomitsu Hotta[1,2]

[1] Research Center for Genetic Engineering and Cell Transplantation; Departments of [2] Internal Medicine, [3] Pediatrics, and [4] Molecular Lifescience, Tokai University School of Medicine, Kanagawa; and [5] Division of Hematology, Pharmaceutical Frontier Research Laboratory, JT Inc., Kanagawa, Japan

Abstract. We established a novel culture system in which the murine stromal cell line HESS-5 dramatically supports the rapid expansion of cryopreserved cord blood primitive progenitor cells (CB-PPC) in synergy with thrombopoietin and Flk-2/Flt-3 ligand. Within 5 days of serum-free culture in this system, a 50- to 100-fold increase in CD34(+)CD38(−) cells was obtained and colony-forming units in culture (CFU-C) and mixed colonies (CFU-GEMM) were amplified by 10- to 30-fold and 10- to 20-fold, respectively. To further assess the long-term repopulating ability of those expanded cells, we performed long-term culture-initiating cells (LTC-IC) assay and SCID-repopulating cells (SRC) assay using CD34(+) cells cultured in this system. Within 5 days of culture 5.1 ± 2.3 fold amplification of the LTC-IC was obtained. SCID-repopulating cells and their multilineage differentiation were detected in NOD/SCID mice 10 weeks after transplantation of 1×10^5 cultured cells. This system is further applicable to retrovirus-mediated gene transfer to CB-PPC. The transduction efficiency of CD34(+) cells was approximately 20–30% when they are infected on HESS-5 monolayer cells. The engraftment of transduced cells in NOD/SCID mice 10 weeks after transplantation was confirmed by the presence of the transduced gene. These results indicate that this xenogeneic coculture system, in combination with the cytokines, can rapidly expand CB stem cells and are applicable to the efficient retrovirus-mediated gene transfer to them.

Keywords: cytokines, gene transfer, hematopoietic system, regeneration, retrovirus, stromal cell line.

Introduction

Hematopoietic stem cell transplantation has been the most effective therapeutic method for hematologic malignancies and bone marrow failures. Recent advances in this field make it possible to use a wide variety of sources for stem cells; bone marrow, peripheral blood, cord blood and fetal liver. Among these, cord blood (CB) is an attractive source because it contains a high number of primitive progenitor cells (PPC) [1,2]. To date, over 700 CB transplantations

Address for correspondence: Kiyoshi Ando MD, Division of Hematology, The Department of Internal Medicine, Tokai University School of Medicine, Bohseidai, Isehara, Kanagawa, 259-1193, Japan. Tel.: +81-463-93-1121 ext. 2230. Fax: +81-463-92-4511. E-mail: andok@keyaki.cc.u-tokai.ac.jp

(CBT) have been performed worldwide and the results over the past 9 years of CBT for children have been promising [3,4]. However, there are potential limitations to the widespread use of CB; there may be enough hematopoietic stem cells to reconstitute children, but not adults. Therefore, ex vivo expansion of PPC is required before CBT for adults.

Many studies to identify culture condition able to support expansion of PPC have been attempted [5—7]. Recently it was reported that the combination of thrombopoietin (TPO) and Flk-2/Flt-3 ligand (FL-2) maintained production of CB-PPC for more than 6 months in stroma-free liquid culture [8]. Both early acting cytokines were reported to sustain cell viability and promote proliferation preferentially of a minor subpopulation of CD34(+) cells, i.e., CD34(+)CD38(–) cells [9—12]; multipotent progenitor cells and long-term culture-initiating cells (LTC-IC) and SCID-repopulating cells (SRC) are included in this subset [13,14].

On the other hand, the coculture system with bone marrow stromal cells was reported to preserve human PPC quality during ex vivo manipulation [15]. We have established several murine stromal cell lines from bone marrow and spleen [16]. Among these, we discovered a novel hematopoietic-supportive cell line, HESS-5, which effectively supports not only formation of murine granulocyte and macrophage colonies but also proliferation of human CD34(+)CD38(–) cells in the presence of human cytokines [17,18]. Progenitors expanded by this xenogeneic coculture system generate a number of high proliferative potential colony-forming cells and mixed colony-forming units and differentiate to CD10+/CD19+ B-lymphoid cells.

We demonstrate here the marked supportive effects of HESS-5 cells on proliferation of PPC isolated from cryopreserved CB in synergy with TPO/FL-2 using a novel culture system by which contamination with murine stromal cells can be avoided. The number of CD34(+)CD38(–) cells was dramatically increased in serum-free condition for 5 days. LTC-IC and SRC assays of these cells revealed their extensive ability to sustain long-term hematopoiesis. This system was further applicable to the efficient retrovirus-mediated gene transfer into CB-PPC.

Materials and Methods

CD34(+) cell purification

CB, collected according to the institutional guidelines, was obtained from normal full-term deliveries. CD34(+) cell purification by using the MACS immunomagnetic separation system (Miltenyi Biotec, Glodbach, Germany) was according to the manufacturer's instructions. Ninety-five percent or more of the enriched cells were CD34(+) by flow cytometric analysis.

Murine stromal cell line

The murine hematopoietic-supportive stromal cell line HESS-5 was previously established from murine bone marrow [16]. HESS-5 cells were maintained in MEM-α supplemented with 10% horse serum at 37°C under 5% CO_2 in humidified air.

Culture systems

Diagrams of several culture systems used in this study are displayed in Fig. 1. Stroma-free culture (A) and classical coculture with stromal cells (B) were performed in culture media in 24-well microplates. In some experiments, human hematopoietic cells were physically separated from the stromal layer by a membrane in cell culture inserts (C). In this study, a novel culture system was devised by modifying the cell culture insert system. First, HESS-5 cells were cultured on

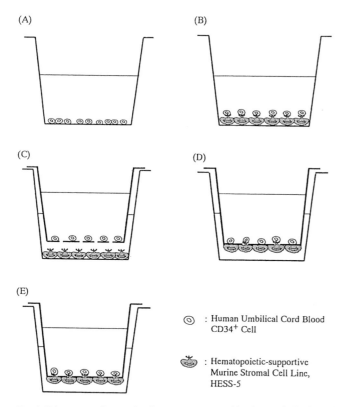

Fig. 1. Diagrams of several culture systems used in the study [26]. **A:** Stroma-free culture; **B:** classical coculture; **C:** noncontact culture; **D:** novel culture using the LPD membrane; **E:** novel culture using the HPD membrane.

the reverse side of the membrane of the insert. After obtaining a confluent monolayer, CD34(+) cells were seeded on the upper side of the membrane of the insert where the cytoplasmic villi of HESS-5 cells passed through the etched 0.45-µm pores. Therefore, while HESS-5 cells directly adhered to human hematopoietic cells during culture, expanded cells could easily be harvested without contamination with HESS-5 cells. Low (D) and high pore-density membranes (E) (LPD, 1.6×10^6/cm^2; HPD, 1.0×10^8/cm^2) were used in the present study.

Short-term ex vivo expansion of hematopoietic progenitors

Serum-containing liquid culture was carried out using a medium containing 12.5% horse serum, 12.5% fetal bovine serum, and MEM-α supplemented with or without designated cytokines. Serum-free liquid culture was carried out using StemProTM-34SFM (GibcoBRL) supplemented with StemProTM-34 nutrient supplement (GibcoBRL) with or without designated cytokines. The final concentrations of cytokines were as follows: TPO, 50 ng/ml; FL-2, 50 ng/ml; IL-3, 20 ng/ml; SCF, 50 ng/ml; GM-CSF, 10 ng/ml; and EPO, 3 U/ml.

LTC-IC assay

LTC-IC assay was performed as described by Sutherland et al. [19] with slight modifications. Briefly, bone marrow stromal cells derived from hematologically normal donors were seeded at 10^5 cells per well in 96-well flat-bottomed plates. CD34(+) cells purified from CB or those isolated from cultured cells were seeded at limiting dilution on the feeder layer. After 5 weeks of culture, cells were assayed for CFU-C in methylcellulose medium.

SCID repopulating cell assay

SRC assay was performed as described by Hogan et al. [20] with slight modifications. Eight-week-old NOD/shi-*scid/scid* (NOD/SCID) mice were obtained from Central Institute for Experimental Animals (Kawasaki, Japan) and maintained in the germ-free animal facility located at Tokai University School of Medicine. Purified cell populations at the indicated dose were transplanted by tail-vein injection into sublethally irradiated mice (350 cGy). Mice were killed 10 weeks after transplantation, and the BM from the femurs of each mouse were flushed into the medium.

Retrovirus-mediated transduction of CD34(+) CB cells

The therapeutic retrovirus vector MFG-GC-GFP and producing cells No. 50 were previously described [21]. CD34(+) CB cells were cultured on HESS-5 monolayer or fibronectin-coated plates (10 µg/cm^2) and infected with virus supernatants supplemented with the cytokines for 4 days.

Statistical analysis

Results are given as mean SEM for three different experiments. Data were compared using analysis of variance. Where significant differences were inferred, sample means were compared using the t test.

Results

Evaluation of supportive effects of HESS-5 cells on ex vivo expansion of CB-PPC in synergy with TPO/FL-2

We first assessed the synergistic effects of HESS-5 cell line with TPO and FL-2 on expansion of CB-PPC in the stroma-free serum-containing culture. As reported, the combination of these factors (TPO/FL-2) enhanced production of progenitors; the number of CD34(+) cells was approximately 4 times the input number after 14 days of culture (Table 1). The supportive effects of HESS-5 cells on proliferation of CB-PPC were then studied. Under cytokine-free conditions, HESS-5 cells did not support proliferation of CD34(+) cells. However, in the presence of TPO/FL-2, HESS-5 cells could effectively support proliferation of progenitor cells, especially in the CD34(+)CD38(–) subpopulation; the mean number of CD34(+)CD38(–) cells was approximately 150 times the input number after 14 days of culture. The output of CFU-C and CFU-Mix also was enhanced.

Table 1. Evaluation of synergistic effects between HESS-5 cells and TPO/FL-2 on ex vivo expansion of hematopoietic progenitors [26].

Cell population or colony	None	HESS-5	TPO/FL-2	HESS-5 & TPO/FL-2
Total number of cells	0.7 ± 0.1	2.9 ± 0.5	98.7 ± 20.7	414.7 ± 10.1
CD34^{bright} cells	0.1 ± 0.0	0.4 ± 0.1	4.3 ± 0.4	16.0 ± 2.5
CD34^{bright}/CD38^{bright} cells	0.1 ± 0.0	0.4 ± 0.1	4.3 ± 0.4	14.7 ± 2.1
CD34^{bright}/CD38^{dim} cells	0.1 ± 0.0	0.4 ± 0.3	4.2 ± 2.0	150.3 ± 84.3^{b}
CFU-C	N.D.	1.7 ± 0.0^{a}	13.3 ± 2.3	74.3 ± 11.2^{c}
CFU-GEMM	N.D.	2.4 ± 0.3^{a}	7.3 ± 5.3	35.7 ± 7.6^{b}

Cord blood (CB) CD34^{bright} cells were cultured in the presence or absence of TPO and FL-2 with or without HESS-5 cells. HESS-5 cells directly adhered to the CB cells through 0.45-μm pores of the low-pore density membrane in the cell culture insert system. Each number at above-mentioned cell populations indicates the mean fold increase ± SEM of three different experiments on day 14 of culture. Suitable aliqouts of cultured cells were assayed for CFU-C and CFU-GEMM. After 2 weeks of the clonal cell culture, colony scoring was performed and the results represent the fold increase (mean ± SEM of three different experiments). ^{a}Only one experiment was performed in triplicate; $^{b}p < 0.05$; $^{c}p < 0.01$ vs. those without HESS-5. N.D.: not determined.

Assessment of the supportive effects of HESS-5 cells on proliferation of CB-PPC using several culture systems

To elucidate the mechanism of supportive effects of HESS-5 cells on the proliferation of CB-PPC, CD34(+) cells were cocultured with HESS-5 cells in several culture systems (Fig. 1). During culture, there were no significant differences in the number of total nucleated cells between the different culture systems (Fig. 2). However, CD34(+) cells, especially CD34(+)CD38(–), were generated more by the HPD culture and the classical coculture than by the LPD culture or the non-

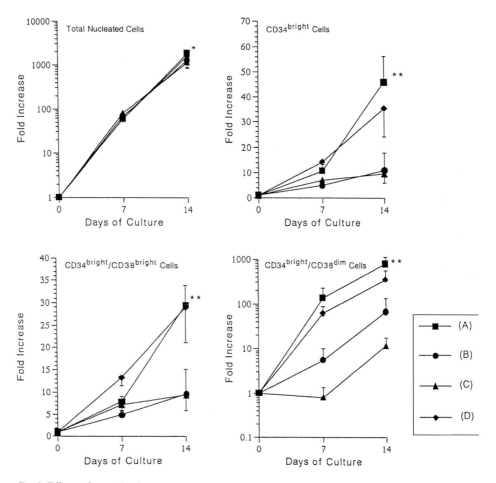

Fig. 2. Effects of several culture systems on ex vivo expansion of hematopoietic progenitors. CD34(+) cells were cocultured with HESS-5 cells in serum-containing medium supplemented with TPO and FL-2 using several culture systems [26]. **A:** HPD culture; **B:** LPD culture; **C:** noncontact culture; **D:** classical coculture. The results represent the mean fold increase ± SEM of three different experiments on days 7 and 14 of culture. *No significant differences between different culture systems ($p > 0.05$). **$p < 0.05$ as compared with LPD culture or noncontact culture.

contact culture; the mean number of CD34(+)CD38(–) cells was over 100 times the initial input number in HPD culture within 7 days. These results indicate that the direct interaction between HESS-5 and CB-PPC rather than humoral factor derived from HESS-5 was required for amplification of CD34(+)CD38(–) cells.

Ex vivo expansion of CB-PPC in a very short-term serum-free culture (VST-SFC)

The synergistic effects of HESS-5 cells and TPO/FL-2 were also studied in serum-free HPD culture for a very short duration. After 5 days of culture, the number of progenitor cells, especially CD34(+)CD38(–) cells, was remarkably increased in the presence of TPO/FL-2 (Table 2). By addition of SCF or IL-3, those effects were further enhanced. As a result, the mean number of CD34(+)CD38(–) cells was approximately 50 to 100 times the initial input number by the VST-SFC system. The outputs of CFU-C and CFU-GEMM were increased 10- to 30-fold and 10- to 20-fold, respectively. Three-color flow cytometry revealed that most of the CD34(+)CD38(–) population amplified by this system was negative for most lineage-committed markers but was positive for HLA-DR (Table 3).

LTC-IC and SRC assay using the CD34(+) population isolated from cells cultured in VST-SFC

To determine whether cells cultured in VST-SFC could preserve the ability to sustain long-term hematopoiesis, quantitation of LTC-IC frequency in vitro and the presence of SRC in vivo was assessed. CD34(+) CB cells cultured in the abovementioned conditions, as well as those initially prepared from CB (control sam-

Table 2. Evaluation of synergistic effects between HESS-5 cells and TPO/FL-2-containing cytokines on ex vivo expansion of hematopoietic progenitors in very short-term serum-free culture [26].

Cell population or colony	Cytokine-free	TPO+FL-2	TPO+FL-2+SCF	TPO+FL-2+IL-3
Total number of cells	1.1 ± 0.0	5.3 ± 0.6	14.3 ± 1.2^a	17.2 ± 3.9^b
CD34^{bright} cells	1.1 ± 0.0	2.3 ± 0.4	4.8 ± 0.1^a	4.9 ± 1.3^a
CD34^{bright}/CD38^{bright} cells	1.1 ± 0.0	1.4 ± 0.2	3.0 ± 0.4^a	2.9 ± 0.8
CD34^{bright}/CD38^{dim} cells	1.1 ± 0.4	48.4 ± 7.6	103.1 ± 10.3^a	113.9 ± 21.7^b
CFU-C	1.5 ± 0.1	12.4 ± 2.3	34.2 ± 6.7^b	33.2 ± 5.3^a
CFU-GEMM	0.5 ± 0.1	9.4 ± 2.5	18.3 ± 5.3	11.5 ± 1.2

In the presence or absence of the above-mentioned cytokines, CB CD34^{bright} cells were cultured with HESS-5 cells under serum-free conditions. HESS-5 cells directly adhered to the CB cells through 0.45-μm pores of the high-pore density membrane in the cell culture insert system. Each number at cell populations indicates the mean fold increase ± SEM of three different experiments on day 5 of culture. Suitable aliquots of cultured cells were assayed for CFU-C and CFU-GEMM. After 2 weeks of the clonal cell culture, colony scoring was performed and the results represent the fold increase (mean ± SEM of three different experiments). ^{a}p < 0.05; ^{b} < 0.01 as compared with TPO+FL-2.

Table 3. Immunophenotype of hematopoietic progenitor cells expanded by the VST-SFC system [26].

Marker	CD34^{bright}/CD38^{bright} cells	CD34^{bright}/CD38^{dim} cells
CD15	29.6 ± 8.2	3.2 ± 1.1
CD19	3.9 ± 2.6	0.0 ± 0.0
CD2	4.8 ± 2.3	0.7 ± 0.7
CD41b	4.7 ± 3.3	6.4 ± 2.4
Gly A	4.5 ± 3.8	2.5 ± 1.6
HLA-DR	88.7 ± 3.8	90.9 ± 3.7

Immunophenotypes of CD34^{bright} cells expanded by the VST-SFC system in the presence of TPO/FL-2 and IL-3 were assessed using three-colour flow cytometry. Each number indicates the mean percentage of positive cells ± SEM of three different experiments.

ples) were used in both assays. The LTC-IC frequency was amplified 5.1 ± 2.3 fold in CD34(+) fraction and 25.3 ± 16.5 fold in CD34(+)CD38(–) fraction when compared with control (Table 4). Seventeen percent of human cells were detected 10 weeks after transplantation of 5×10^5 CD34(+) cells into NOD/SCID and these cells were differentiated into B lymphocytes and myeloid lineage (Fig. 3).

Retroviral transduction of CD34(+) and SRC by using HESS-5 coculture system

For the aim of gene therapy of Gaucher disease we have constructed a bicistronic retrovirus vector (pMFG-GC-GFP) which contains both the human glucocerebrosidase (GC) and the green fluorescent protein (GFP) gene as a selectable marker [21]. The transduction efficiency of human CB CD34(+) cells was approximately 20–30% by either the HESS-5 or the FN method (Fig. 4). The transduced cells also confirmed the marked increase in enzyme activity of GC (data not shown). Irradiated NOD/SCID mice were injected with 1×10^5 transduced CD34(+) cells, and sacrificed after 10 weeks. Fifteen percent of human CD45(+) cells were detected in mice transplanted with cells transduced on HESS-5. The transduced cells confirmed the presence of human GC cDNA by

Table 4. Frequency and fold increase of LTC-IC by the VST-SFC system [26].

	Frequency of LTC-IC		Total fold increase
	Pre-expansion	Postexpansion	
CD34+	1/102	1/138	5.1 ± 2.3
CD34+38+	1/215	1/2833	0.17 ± 0.1
CD34+38–	1/88	1/134	25.3 ± 16.5

Same number of CB cells before or after culture in the VST-SFC system were seeded on the human bone marrow stromal cells in 96-well plates. After 5 weeks of culture, cells were assayed for CFU-C, the frequency of wells in which there were no clonogenic cells were determined per the numbers of initial input populations and the frequency of LTC-IC was calculated. Total fold increase was shown by mean ± SEM of three different experiments.

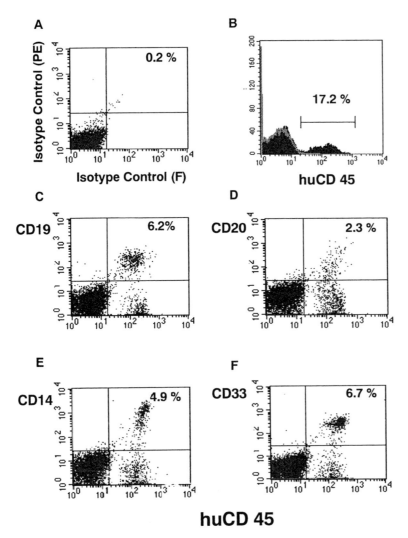

huCD 45

Fig. 3. Engraftment and multilineage differentiation of ex vivo expanded CB CD34(+) cells in NOD/ SCID mice. Bone marrow from a highly engrafted mouse transplanted with 1×10^5 CD34(+) cells was stained with various human-specific monoclonal antibodies and analyzed by flow cytometry. **A:** Isotype control of nonspecific IgG staining of PE and FITC fluorescence. **B:** Histogram of human CD45 expression indicating that 17.2% of the cells present in the murine bone marrow are human. Analysis of lineage markers was done on cells within CD45(+). Populations of (**C,D**) B lymphocytes (CD19(+), 6.2%, CD20(+), 2.3%), monocytes (**E**) (CD14(+), 4.9%), and myeloid progenitors (CD33(+), 6.7%) were observed.

PCR and Southern blotting. In contrast, less than 1% of human cells were detected in mice transplanted with cells transduced in fibronectin-coated plates (Fig. 5). These results indicate that the HESS-5 supports the efficient retrovirus-mediated gene transfer to human hematopoietic stem cells.

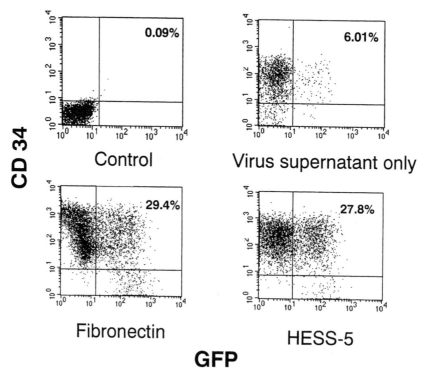

GFP

Fig. 4. Retrovirus-mediated gene transfer into CD34(+) CB cells [27]. CD34(+) CB cells were cultured in 6-well plate and infected with 3 ml of virus supernatants (1×10^6 pfu/ml) supplemented with cytokines for 4 days in various conditions. Expression of GFP was analyzed by flow cytometry. Control; isotype control of nonspecific IgG staining of PE and FITC fluorescence. Fibronectin; cells were infected on fibronectin-coated plate. HESS-5; cells were infected on HESS-5 monolayer.

Discussion

In this study we demonstrated that the murine stromal cell line HESS-5 could effectively support rapid expansion of cryopreserved CB-PPC in synergy with human cytokines. While HESS-5 cells alone did not effectively support proliferation, HESS-5 cells could dramatically enhance generation of progenitors, especially CD34(+)CD38(–) cells, in synergy with TPO/FL-2 over a short period. These expanded cells were confirmed to sustain long-term repopulation in LTC-IC and SRC assays. This system was further applicable to retrovirus-mediated gene transfer into CB stem cells.

In ex vivo expansion of human PPC, there are several benefits to using a xenogeneic stromal cell line as the feeder: 1) HESS-5 cells can be maintained easily, 2) consistent hematopoietic-supportive effects are repeatedly obtained, and 3) the exposure of PPC to differentiation-inducing factors acting on them derived from stromal cells can be avoided. We previously studied the supportive effects of normal human bone marrow fibroblasts on proliferation of CB-PPC, but vari-

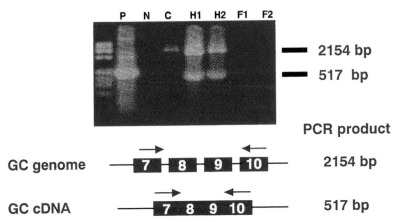

Fig. 5. Detection of the transduced GC cDNA in SRC [27]. High molecular DNA was isolated from the bone marrow cells from NOD/SCID mice injected with 1×10^5 CD34(+) CB cells and analyzed using the PCR and Southern blotting with a human GC cDNA probe. Sense primer in exon7 and antisense primer in exon10 were used for PCR so that the fragments from genomic GC (2154 bp) and cDNA (517bp) can be discriminated. P; MFG-GC-GFP. N; no DNA. C; DNA from a mouse transplanted with 1×10^4 CD34(+) CB cells. H1 and H2; two different mice transplanted with 1×10^5 transduced CD34(+) CB cells on HESS-5 cells. F1 and F2; two different mice transplanted with 1×10^5 transduced CD34(+) CB cells on fibronectin-coated plate.

able results were noted because of their differentiation. (data not shown). However, a potential problem is contamination of xenogeneic stromal cells into cultured human progenitors. Therefore, we established a novel culture system in which HESS-5 cells were easily seeded and grew on the reverse side of the membrane of the cell culture insert; cultured human cells were simply harvested without contamination by HESS-5 cells. Evaluation of the hematopoietic-supportive effects of HESS-5 cells using several different culture systems suggested that proliferation of CB-PPC enhanced by HESS-5 cells was not due to soluble factors produced by HESS-5 cells but was mediated by direct cell-to-cell interaction. Therefore, the HPD culture was more effective for expansion of PPC than LPD culture, indicating that progenitor-stroma-contact density is important for effective expansion.

On the basis of these observations, we further assessed the HPD culture system in serum-free conditions for a very short duration and adequate expansion of CB-PPC was obtained in this system. Most of the expanded CD34(+)CD38(–) cell population expressed HLA-DR. In contrast to previous findings in bone marrow, CB LTC-IC were shown to be present among the CD34(+)HLA-DR(+) cell fraction [22]. Therefore, the expanded hematopoietic progenitors were expected to sustain long-term hematopoiesis. As a result, the LTC-IC number was increased up to 10-fold within 5 days and SRC was confirmed in the expanded fraction of cells. These results open the gate to the application of ex vivo expansion of hematopoietic stem cells to CBT for adults. The safety issues of using the xenogeneic system in clinical setting are now under investigation.

Although hematopoietic stem cells are ideal targets for gene therapy, the transduction efficiency has been too low to use clinically [23]. As shown above, however, the application of coculture system with HESS-5 enabled the high level of gene transfer into CD34(+) CB cells with SRC ability. These results indicate that the xenogeneic coculture system mentioned here, in combination with human cytokines, can rapidly expand CB stem cells and is applicable to the efficient retrovirus-mediated gene transfer to them.

The mechanism of hematopoietic-supportive effects of HESS-5 cells, especially in the CD34(+)CD38(−) cell fraction, remains unknown. Recently, cell surface molecules identified as δ-like/preadipocyte factor-1 or jagged-1 on stromal cells were reported to show activity of preserving the primitive phenotype of hematopoietic progenitors [24,25]. These novel molecular pathways may play an important role in stem cell regulation in the system we presented here.

Acknowledgements

The authors thank the members of Tokai CBSC Study Group for providing human cord blood. This study was supported by The Japan Society for the Promotion of Science (JSPS) grant No. JSPS-RFTF97I00201 and a Research Grant of The Science Frontier Program from the Ministry of Education, Science, Sports and Culture of Japan.

References

1. Nakahata T, Ogawa M. Hematopoietic colony-forming cells in umbilical cord blood with extensive capability to generate mono- and multipotential hemopoietic progenitors. J Clin Invest 1982;70:1324.
2. Broxmeyer HE, Douglas GW, Hangoc G, Cooper S, Bard J, English D, Arny M, Thomas L, Boyse EA. Human UCB as a potential source of transplantable stem/progenitor cells. Proc Natl Acad Sci USA 1989;86:3828.
3. Cairo MS, Wagner JE. Placental and/or umbilical cord blood: an alternative source of hematopoietic stem cells for transplantation. Blood 1997;90:4665.
4. Gluckman E, Rocha V, Boyer-Chammard A, Locatelli F, Arcese W, Pasquini R, Ortega J, Souillet G, Ferreira E, Laporte JP, Fernandez M, Chastang C. Outcome of cord-blood transplantation from related and unrelated donors. Eurocord Transplant Group and the European Blood and Marrow Transplantation Group. N Engl J Med 1997;337:373.
5. Emerson S. Ex vivo expansion of hematopoietic precursors, progenitors, and stem cells: The next generation of cellular therapeutics. Blood 1995;85:2059.
6. Sui X, Tsuji K, Tanaka R, Tajima S, Muraoka K, Ebihara Y, Ikebuchi K, Yasukawa K, Taga T, Kishimoto T, Nakahata T. gp130 and c-Kit signalings synergize for ex vivo expansion of human primitive hemopoietic progenitor cells. Proc Natl Acad Sci USA 1995;92:2859.
7. Petzer AL, Hogge DE, Landsdorp PM, Reid DS, Eaves CJ. Self-renewal of primitive human hematopoietic cells (long-term-culture-initiating cells) in vitro and their expansion in defined medium. Proc Natl Acad Sci USA 1996;93:1470.
8. Piacibello W, Sanavio F, Garetto L, Severino A, Bergandi D, Ferrario J, Fagioli F, Berger M, Aglietta M. Extensive amplification and self-renewal of human primitive hematopoietic stem cells from cord blood. Blood 1997;89:2644.

9. Petzer AL, Zandstra PW, Piret JM, Eaves CJ. Differential cytokine effects on primitive (CD34+CD38–) human hematopoietic cells: novel responses to Flt3-ligand and thrombopoietin. J Exp Med 1996;183:2551.

10. Kobayashi M, Laver JH, Kato T, Miyazaki H, Ogawa M. Thrombopoietin supports proliferation of human primitive hematopoietic cells in synergy with steel factor and/or interleukin-3. Blood 1996;88:429.

11. Haylock DN, Horsfall MJ, Dowse TL, Ramshaw HS, Niutta S, Protopsaltis S, Peng L, Burrell C, Rappold I, Buhring H-J, Simmons PJ. Increased recruitment of hematopoietic progenitor cells underlies the ex vivo expansion potential of FLT3 ligand. Blood 1997;90:2260.

12. Borge OJ, Ramsfjell V, Cui L, Jacobsen SEW. Ability of early acting cytokines to directly promote survival and suppress apoptosis of human primitive CD34+CD38– bone marrow cells with multilineage potential at the single-cell level: key role of thrombopoietin. Blood 1997; 90:2282.

13. Hao QL, Shah AJ, Thiemann FT, Smogorzewska EM, Crooks GM. A functional comparison of CD34+CD38– cells in cord blood and bone marrow. Blood 1995;86:3745.

14. Bhatia M, Bonnet D, Kapp U, Wang JCY, Murdoch B, Dick JE. Quantitative analysis reveals expansion of human hematopoietic repopulating cells after short-term ex vivo culture. J Exp Med 1997;186:619.

15. Breems DA, Blokland EAW, Siebel KE, Mayen AEM, Engels LJA, Ploemacher RE. Stroma-contact prevents loss of hematopoietic stem cell quality during ex vivo expansion of CD34+ mobilized peripheral blood stem cells. Blood 1998;91:111.

16. Tsuji T, Ogasawara H, Aoki Y, Tsurumaki Y, Kodama H. Characterization of murine stromal cell clones established from bone marrow and spleen. Leukemia 1996;10:803.

17. Tsuji T, Watanabe Y, Nishimura Y, Kamada M. Dramatic short-term expansion of CD34high+CD38– primitive progenitor cells isolated from human umbilical cord blood in a xenogeneic coculture system in combination with human IL-3 and SCF. Blood 1997;90:366a (Abstract).

18. Tsuji T, Watanabe Y, Nishimura Y, Waga I, Yatsunami K. In vitro expansion of CD34high+ cells isolated from human umbilical cord blood on a murine stromal cell line in combination with human cytokines. Blood 1997;90:366a (Abstract).

19. Sutherland HJ, Lansdorp PM, Henkelman DH, Eaves AC, Eaves CJ. Functional characterization of individual human hematopoietic stem cells cultured at limiting dilution on supportive marrow stromal layers. Proc Natl Acad Sci USA 1990;87:3584.

20. Hogan CJ, Shpall EJ, McNulty O, Dick JE, Shultz LD, Keller G. Engraftment and development of human CD34(+)-enriched cells from umbilical cord blood in NOD/LtSz-scid/scid mice. Blood 1997;90:85.

21. Shimizu T, Ando K, Kimura M, Miyatake H, Inokuchi S, Takakura I, Migita M, Shimada T, Kato S. A simple and efficient purification of transduced cells by using green fluorescent protein gene as a selection marker. Acta Pediatrica Japonica 1998;40:575.

22. Traycoff CM, Abboud MR, Laver J, Brandt JE, Hoffman R, Law P, Ishizawa L, Srour EF. Evaluation of the in vitro behavior of phenotypically defined populations of umbilical cord blood hematopoietic progenitor cells. Exp Hematol 1994;22:215.

23. Larochelle A, Vormoor J, Hanenberg H, Wang JCY, Bhatia M, Lapidot T, Moritz T, Murdoch B, Xiao XL, Kato I, Williams DA, Dick JE. Identification of primitive human hematopoietic cells capable of repopulating NOD/SCID mouse bone marrow: implications for gene therapy. Nat Med 1996;2:1329.

24. Moore KA, Pytowski B, Witte L, Hicklin D, Lemischka IR. Hematopoietic activity of a stromal cell transmembrane protein containing epidermal growth factor-like repeat motifs. Proc Natl Acad Sci USA 1997;94:4011.

25. Varnum-Finney B, Purton LE, Yu M, Brashem-Stein C, Flwers D, Staats S, Moore KA, Le Roux I, Mann R, Gray G, Artavanis-Tsakonas S, Bernstein ID. The Notch ligand, jagged-1, influences the development of primitive hematopoietic precursor cells. Blood 1998;91:4084.

26. Kawada H, Ando K, Tsuji T, Shimakura Y, Nakamura Y, Chargui J, Hagihara M, Itagaki H, Shimizu T, Inokuchi S, Kato S, Hotta T. Rapid ex vivo expansion of human umbilical cord hematopoietic progenitors using a novel culture system. Exp Hematol 1999;(In press).
27. Shimizu T, Ando K, Tsuji T, Kimura M, Miyatake H, Sato T, Nakamura Y, Hotta T, Kato S. Efficient retrovirus-mediated gene transfer to human CB-SRC using HESS-5 stromal cell line. Blood 1998;92:470a.

Tissue Engineering for Therapeutic Use 3.
Y. Ikada and T. Okano, editors.

Expansion of hematopoietic stem cells — new possibilities

C. Eaves[1,2], E. Conneally[1,3], J. Audet[4,6], S. Rose-John[7], A. Eaves[1,3,5] and J. Piret[4,6]

[1] *Terry Fox Laboratory, British Columbia Cancer Agency; Departments of* [2] *Medical Genetics,* [3] *Pathology and Laboratory Medicine,* [4] *Chemical Engineering,* [5] *Medicine, and* [6] *The Biotechnology Laboratory, University of British Columbia, Vancouver, British Columbia, Canada; and* [7] *I. Medizinische Klinik, Johannes Gutenberg-Universitat, Mainz, Germany*

Abstract. This review summarizes some of the evidence underlying a hierarchical model of early hematopoietic cell differentiation events in both human and murine systems. It highlights the key features of currently available functional assays (both in vitro and in vivo) for quantitating their most primitive elements. These assays have been used to identify and validate the components of culture media (particularly the types and concentrations of the growth factors added) that influence whether the relevant functional properties of these cells are retained or altered when they are stimulated to proliferate in vitro. Conditions that support modest but consistent expansions of these functionally defined hematopoietic stem cells have already been found. Additionally, conditions that do not alter the survival or proliferative response of these cells, and also do not support the maintenance of their stem cell functions have been discovered. Finally, a new level of complexity in terms of how hematopoietic stem cells are regulated has emerged from analyses of cells from ontologically different sources. These findings help to explain previous difficulties in achieving dramatic in vitro expansions of very primitive hematopoietic cells in spite of early excitement based on studies of clonogenic progenitors.

Keywords: blood progenitors, bone marrow transplantation, growth factors, self-renewal.

Introduction

The presence of persistent, dominant clones of blood cells including both lymphoid and myeloid elements appears to be a relatively rare situation in normal individuals [1]. However, in occasional patients recovering from myeloablative doses of radio- and/or chemotherapy after receiving a transplant of either autologous or allogeneic marrow cells, such clones have been documented using a variety of genetic strategies [2,3]. Calculation of the accumulated size of these clones (based on known kinetics of normal human granulocytes) indicates that the proliferative potential of some human hematopoietic stem cells must exceed 40 divisions (to produce $> 10^{13}$ progeny). Unfortunately, the development of procedures for harnessing this proliferative activity to achieve the large-scale expansion ex vivo of human hematopoietic stem cells has been much slower and more

Address for correspondence: Dr C.J. Eaves, Terry Fox Laboratory/BC Cancer Agency, 601 West 10th Avenue, Vancouver, BC V5Z 1L3, Canada. Tel.: +1-604-877-6070 ext. 3146. Fax: +1-604-877-0712. E-mail: connie@terryfox.ubc.ca

difficult. In part, this reflects the complexity of very primitive hematopoietic cell types as well as their regulatory mechanisms, many of which still remain to be clarified. Nevertheless, the last few years has seen a number of significant advances in this area. In this article, progress emanating from recent findings that our group have contributed is reviewed.

A recurrent theme in these studies is the dependence of the interpretation of the results on the assay used to detect stem cell properties and at the same time quantitate their numbers. For the purposes of clinical transplants, the defining property of interest is the capacity to rapidly repopulate the hematopoietic system after intravenous infusion and then continue to meet the lifelong blood supply needs of the recipient. Clearly only clinical studies can allow definitive characterization of such cells and, in the interim, reliance on surrogate assays is required. An early candidate now widely used for this purpose is the long-term culture-initiating cell (LTC-IC). LTC-ICs were first defined as cells that can give rise to clonogenic hematopoietic (myeloid) cells after $\geqslant 4$ (mouse [4,5]) or $\geqslant 5$ (human [6]) weeks in cultures containing pre-established layers of irradiated adherent marrow stromal cells. This definition was based on observations that such clonogenic cell precursors were different from the majority of directly clonogenic cells (detectable by their ability to produce colonies of one or more lineages of mature myeloid cells in semisolid medium). Moreover, in the mouse, LTC-ICs copurify with long-term repopulating cells. Subsequent studies have shown that a variety of immortalized mouse fibroblast cell lines can effectively substitute for the primary marrow stromal layers originally used [7–10] and that a higher plating efficiency of human LTC-ICs can be achieved if certain growth factors are also added or their endogenous production within the culture is engineered [10,11]. In addition, the critical importance of the duration of the culture prior to assessing its colony-forming cell (CFC) content has resulted in the now preferred use of longer assays (routinely 6 weeks in our experiments since 1996) to identify LTC-ICs that most closely correlate with long-term in vivo repopulating cells [11,12].

Expansion of human LTC-IC numbers in vitro

Isolation of the tiny subpopulation ($\leqslant 0.1\%$) of cells in normal adult human marrow that express high levels of the CD34 antigen on their surface but no detectable CD38 yields a suspension that is highly enriched in LTC-IC [13,14]. When these $CD34^+CD38^-$ cells are cultured at dilute concentrations in serum-free medium containing certain growth factors for 10–30 days, LTC-IC expansions of >50-fold occur [14–16]. Systematic factorial design experiments have revealed that, of a large number of growth factors surveyed, flt3-ligand (FL), steel factor (SF) and interleukin-3 (IL-3) are necessary and sufficient to achieve this expansion [17]. In addition, both the absolute and relative concentrations of these factors are important [15]. The expansion of LTC-ICs that takes place is accompanied by a much larger increase in the total number of cells produced, most of which after 10 days still have CFC activity. Expansion of this latter population can be increased further only if one of several other growth factors is also added [17], but is less sensitive to changes in either the absolute or relative concentrations of the added growth factors [15]. Moreover, single cell experiments have indicated that the biological properties of proliferating LTC-ICs can be differentially affected by the growth factor conditions used to stimulate them resulting, alternatively, in the retention or loss of their LTC-IC status [15]. These findings suggest that the type and intensity of intracellular signalling pathways activated in very primitive hematopoietic cells can either force or delay their differentiation. The ability to regulate this decision by controlling the extracellular milieu in which a hematopoietic stem cell is contained opens important possibilities while at the same time posing additional engineering challenges to the design and construction of suitable bioreactor systems.

Interestingly, analogous studies of primitive human hematopoietic cells isolated from cord blood samples have revealed another dimension to the molecular control of LTC-IC and CFC amplification from $CD34^+CD38^-$ cells [16]. This is manifested as a change during ontogeny in some of the cytokines that are able to regulate these responses, as summarized in Fig. 1. Of particular note is the apparent importance of gp 130 activation for stimulating LTC-IC amplification from human cord blood, but not adult bone marrow precursors; consistent with previous studies of human cord blood cell responses [18,19]. More recently, we have confirmed (Table 1) and extended these observations using a novel recombinant fusion protein that combines human IL-6 and the soluble human IL-6 receptor (sIL-6R) via a flexible linker to yield a human gp 130 agonist with elevated potency [20] (Table 2). This engineered cytokine has thus been called hyper-IL-6 (H-IL-6). The data summarized in Tables 1 and 2 show that FL and

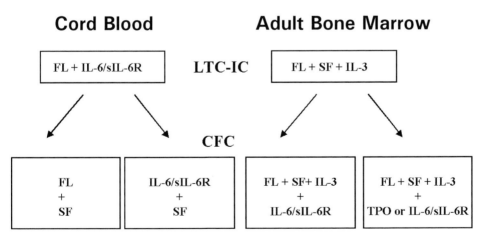

Fig. 1. Comparison of the different cytokine requirements for optimizing the amplification of LTC-IC and various types of CFC in 10-day cultures of $CD34^+CD38^-$ cells isolated from two ontologically distinct human sources. Adapted from [16].

Table 1. Effects of selected cytokine combinations on the 10-day expansion of CFC and LTC-IC populations from purified CD34$^+$CD38$^-$ human cord blood cells.

Cytokine combination	CFC expansion	LTC-IC expansion
FL, IL-6/sIL-6R	49 ± 32	26 ± 19
FL, TPO	23 ± 9	3 ± 0
FL, SF, IL-6, IL-3, G-CSF	990 ± 70	1 ± 0.3

Viable CD34$^+$CD38$^-$ cells were isolated by fluorescent-activated cell sorting from the light density fraction of human cord blood cells and cultured for 10 days in serum-free medium as previously described [16]. The results show the mean ± SEM of the expansions measured in three independent experiments. The final concentration of the various recombinant human cytokines in the medium at the beginning of the culture were as follows: 300 ng/ml FL, 60 ng/ml IL-6, 1000 ng/ml sIL-6R, 50 ng/ml thrombopoietin (TPO), 300 ng/ml SF, 60 ng/ml IL-3, and 60 ng/ml granulocyte colony-stimulating factor (G-CSF).

SF in combination with H-IL-6 are comparable to FL, SF and IL-6/sIL-6R and either of these combinations stimulate a greater amplification of cord blood LTC-IC than is obtained with the cytokine cocktail identified as optimal for expanding adult marrow LTC-IC populations [16].

Expansions of murine repopulating stem cells

An alternative to the use of surrogate assays for human in vivo hematopoietic stem cell function has been to investigate the murine model where suitable direct assays have been available for a number of years. Following the realization that the spleen colony assay [21] does not have sufficient specificity to be useful for this purpose [22,23] (although clearly some long-term repopulating cells can generate macroscopically visible spleen colonies within 2 weeks [24]), we devised a method to allow limiting dilution analysis to be used to specifically quantitate murine cells with long-term multilineage reconstituting activity [25]. The ≥ 4 month period of engraftment used to establish the durability of their hematopoietic activity is longer than the 1–2 weeks required to detect CFU-S [26–28] and this prolonged endpoint necessitates the separate provision of enough hematopoietic cells to ensure that all recipients will survive independent of the cells

Table 2. Numbers of CFC in direct assays and 6-week LTC-IC assays of cells produced by purified CD34$^+$CD38$^-$ human cord blood cells in 7-day cultures.

Cytokine combination	Direct CFC assays	LTC-IC assays
FL, SF & IL-6/sIL-6R	26000 ± 11400	4870 ± 4330
FL, SF & H-IL-6	22300 ± 8400	2840 ± 1780

Cytokine concentrations are the same as in the footnote to Table 1. H-IL-6 where present was added to give a final initial concentration of 10 ng/ml. Values shown are the mean ± SEM of data obtained in four experiments normalized per 1,000 input cells.

being assayed. In addition, evidence of the production of short-lived myeloid as well as long-lived lymphoid progeny is used to establish the totipotency of the cells injected. The cells thus defined and quantitated are referred to as competitive repopulating units (CRU) [25,29,30].

This assay has made it feasible to show that the murine CRU population can expand significantly in vivo both during normal ontogeny and post-transplant [27,31,32]. Nevertheless, attempts to achieve an equivalent net expansion of CRU numbers in vitro, even under conditions where extensive amplification of some CRU could be demonstrated by retroviral marking [33] consistently failed. More recent appreciation of the potential importance of FL stimulation and gp 130 activation, independent of a cell's competence to respond to IL-6, prompted us to examine the potential effects of a cytokine combination consisting of FL, SF and IL-11 (another cytokine that signals via gp 130) on murine CRU in vitro. These studies showed that a modest (on average, less than 10-fold) but, nevertheless, significant net expansion of both LTC-IC and CRU can be obtained when Sca-1$^+$lin$^-$ adult mouse bone marrow cells are cultured for 10 days in a serum-free medium containing FL, SF, and IL-11 [30]. Importantly, the self-renewal activity of the in vitro amplified CRU was found to be undiminished. More recent experiments have shown that H-IL-6 can effectively substitute for IL-11 [34].

Expansion of human repopulating stem cells

The remarkable ability of intravenously injected human cells to initiate and sustain the production of multiple lineages of human hematopoiesis for many months in the host tissue of adequately immunocompromised xenogeneic recipients is now well-established [35−37]. When these are sublethally irradiated NOD/SCID mice, human B-lymphoid cells as well as multiple lineages of human myeloid cells and their progenitors are all represented among the regenerated populations, even when the original human cells transplanted are injected at limiting dilutions and no human growth factors are provided [38,39]. It has therefore been possible to use this xenotransplant model to develop a quantitative assay for human hematopoietic cells that meet the criteria of a CRU. This has introduced a new dimension to the investigation of the effects of various types of cell culture conditions on the maintenance of human hematopoietic stem cell functions. Initial results to date have demonstrated that a limited expansion of human CRU isolated from cord blood is possible [38,40], even under conditions that we now suspect would be suboptimal. Consistent with this finding has been the ability to use these conditions to design a gene transfer protocol for use with retroviral vectors that gives higher efficiencies ($> 10\%$) of gene transfer to human CRU [12] than have previously been obtainable [41].

Continued investigation of how human CRU functions are regulated, improvements in their detection, and an increased understanding of their properties and heterogeneity will clearly be key to the identification and verification of in vitro conditions that will allow greater expansions of their numbers to be obtained in the future. On the other hand, now that a more direct assay for transplantable human hematopoietic cells is possible, the marked similarities revealed between these cells, their murine counterparts, and human LTC-IC, measured in 6-week CFC output assays, also serve as an important and reassuring reminder of the value of using multiple systems to approach the still-elusive objective of mimicking in vitro at least the extent of hematopoietic stem cell expansion that occurs in vivo.

Summary

The blood-forming system maintains a small population of totipotent hematopoietic stem cells throughout adult life. These stem cells reside in the bone marrow and proliferate at a slow rate. However, when transplanted into myeloablated recipients, they can be stimulated to generate large stable clones containing all myeloid and lymphoid lineages, as well as progeny with the same reconstituting activities. These properties have allowed the development of quantitative functional assays for hematopoietic stem cells from both murine and human sources based on the use of appropriate congenic and xenogeneic transplant models in conjunction with limiting dilution analysis techniques. Recent applications of these assays to stem-cell-containing populations manipulated in vitro and in vivo have identified specific cytokines that stimulate their proliferation and amplification. In addition, evidence that these two responses are not necessarily coordinately regulated has been obtained. These findings lay the basis of new opportunities for producing expanded populations of hematopoietic stem cells and their various progeny for a variety of clinical and gene therapy applications.

Acknowledgements

Much of the work described in this review was supported by grants from the National Cancer Institute of Canada (NCIC) with funds from the Terry Fox Run, from the National Institutes of Health of the USA (POI-HL55435) and from Novartis Canada. C. Eaves is a Terry Fox Cancer Research Scientist of the NCIC. E. Conneally held an NCIC Terry Fox Physician-Scientist Fellowship and J. Audet held a Studentship from the Natural Sciences and Engineering Research Council of Canada and a Scholarship from the Science Council of BC.

References

1. Raskind WH, Fialkow PJ. The use of cell markers in the study of human hematopoietic neoplasia. Adv Cancer Res 1987;49:127–167.
2. Turhan AG, Humphries RK, Phillips GL, Eaves AC, Eaves CJ. Clonal hematopoiesis demonstrated by X-linked DNA polymorphisms after allogeneic bone marrow transplantation. N

Engl J Med 1989;320:1655—1661.

3. Eaves CJ, Shaw G, Horsman D, Barnett MJ, Eaves AC. Long-term oligoclonal hematopoiesis after transplantation of culture-purged autologous bone marrow in chronic myeloid leukemia (CML). Blood 1992;80(1):67a (Abstract).

4. Ploemacher RE, Van Der Sluijs JP, van Beurden CAJ, Baert MRM, Chan PL. Use of limiting-dilution type long-term marrow cultures in frequency analysis of marrow-repopulating and spleen colony-forming hematopoietic stem cells in the mouse. Blood 1991;78:2527—2533.

5. Lemieux ME, Rebel VI, Lansdorp PM, Eaves CJ. Characterization and purification of a primitive hematopoietic cell type in adult mouse marrow capable of lympho-myeloid differentiation in long-term marrow "switch" cultures. Blood 1995;86:1339—1347.

6. Sutherland HJ, Lansdorp PM, Henkelman DH, Eaves AC, Eaves CJ. Functional characterization of individual human hematopoietic stem cells cultured at limiting dilution on supportive marrow stromal layers. Proc Natl Acad Sci USA 1990;87:3584—3588.

7. Sutherland HJ, Eaves CJ, Lansdorp PM, Thacker JD, Hogge DE. Differential regulation of primitive human hematopoietic cells in long-term cultures maintained on genetically engineered murine stromal cells. Blood 1991;78:666—672.

8. Baum CM, Weissman IL, Tsukamoto AS, Buckle AM, Peault B. Isolation of a candidate human hematopoietic stem-cell population. Proc Natl Acad Sci USA 1992;89:2804—2808.

9. Neben S, Anklesaria P, Greenberger J, Mauch P. Quantitation of murine hematopoietic stem cells in vitro by limiting dilution analysis of cobblestone area formation on a clonal stromal cell line. Exp Hematol 1993;21:438—443.

10. Hogge DE, Lansdorp PM, Reid D, Gerhard B, Eaves CJ. Enhanced detection, maintenance and differentiation of primitive human hematopoietic cells in cultures containing murine fibroblasts engineered to produce human steel factor, interleukin-3 and granulocyte colony-stimulating factor. Blood 1996;88:3765—3773.

11. Hao QL, Thiemann FT, Petersen D, Smogorzewska EM, Crooks GM. Extended long-term culture reveals a highly quiescent and primitive human hematopoietic progenitor population. Blood 1996;88:3306—3313.

12. Conneally E, Eaves CJ, Humphries RK. Efficient retroviral-mediated gene transfer to human cord blood stem cells with in vivo repopulating potential. Blood 1998;91:3487—3493.

13. Issaad C, Croisille L, Katz A, Vainchenker W, Coulombel L. A murine stromal cell line allows the proliferation of very primitive human $CD34^{++}/CD38^-$ progenitor cells in long-term cultures and semisolid assays. Blood 1993;81:2916—2924.

14. Petzer AL, Hogge DE, Lansdorp PM, Reid DS, Eaves CJ. Self-renewal of primitive human hematopoietic cells (long-term culture-initiating cells) in vitro and their expansion in defined medium. Proc Natl Acad Sci USA 1996;93:1470—1474.

15. Zandstra PW, Conneally E, Petzer AL, Piret JM, Eaves CJ. Cytokine manipulation of primitive human hematopoietic cell self-renewal. Proc Natl Acad Sci USA 1997;94:4698—4703.

16. Zandstra PW, Conneally E, Piret JM, Eaves CJ. Ontogeny-associated changes in the cytokine responses of primitive human haematopoietic cells. Br J Haematol 1998;101:770—778.

17. Petzer AL, Zandstra PW, Piret JM, Eaves CJ. Differential cytokine effects on primitive ($CD34^+CD38^-$) human hematopoietic cells: novel responses to flt3-ligand and thrombopoietin. J Exp Med 1996;183:2551—2558.

18. Sui X, Tsuji K, Tanaka R, Tajima S, Muraoka K, Ebihara Y et al. gp 130 and c-kit signalings synergize for ex vivo expansion of human primitive hemopoietic progenitor cells. Proc Natl Acad Sci USA 1995;92:2859—2863.

19. Tajima S, Tsuji K, Ebihara Y, Sui X, Tanaka R, Muraoka K et al. Analysis of interleukin 6 receptor and gp130 expressions and proliferative capability of human CD34+ cells. J Exp Med 1996;184:1357—1364.

20. Fischer M, Goldschmitt J, Peschel C, Brakenhoff JPG, Kallen KJ, Wollmer A et al. A bioactive designer cytokine for human hematopoietic progenitor cell expansion. Nature Biotechnol 1997;15:142—145.

21. Till JE, McCulloch EA. A direct measurement of the radiation sensitivity of normal mouse bone marrow cells. Radiat Res 1961;14:213−222.
22. Jones RJ,Wagner JE, Celano P, Zicha MS, Sharkis SJ. Separation of pluripotent haematopoietic stem cells from spleen colony-forming cells. Nature 1990;347:188−189.
23. Ploemacher RE, Brons NHC. Isolation of hemopoietic stem cell subsets from murine bone marrow: II. Evidence for an early precursor of day-12 CFU-S and cells associated with radioprotective ability. Exp Hematol 1988;16:27−32.
24. Dumenil D, Jacquemin-Sablon H, Neel H, Frindel E, Dautry F. Mock retroviral infection alters the developmental potential of murine bone marrow stem cells. Molec Cell Biol 1989;9: 4541−4544.
25. Szilvassy SJ, Humphries RK, Lansdorp PM, Eaves AC, Eaves CJ. Quantitative assay for totipotent reconstituting hematopoietic stem cells by a competitive repopulation strategy. Proc Natl Acad Sci USA 1990;87:8736−8740.
26. Jordan CT, Lemischka IR. Clonal and systemic analysis of long-term hematopoiesis in the mouse. Genes Dev 1990;4:220−232.
27. Miller CL, Rebel VI, Helgason CD, Lansdorp PM, Eaves CJ. Impaired steel factor responsiveness differentially affects the detection and long-term maintenance of fetal liver hematopoietic stem cells in vivo. Blood 1997;89:1214−1223.
28. Morrison SJ,Weissman IL. The long-term repopulating subset of hematopoietic stem cells is deterministic and isolatable by phenotype. Immunity 1994;1:661−673.
29. Rebel VI, Dragowska W, Eaves CJ, Humphries RK, Lansdorp PM. Amplification of Sca-1$^+$ Lin$^-$ WGA$^+$ cells in serum-free cultures containing steel factor, interleukin-6, and erythropoietin with maintenance of cells with long-term in vivo reconstituting potential. Blood 1994;83:128−136.
30. Miller CL, Eaves CJ. Expansion in vitro of adult murine hematopoietic stem cells with transplantable lympho-myeloid reconstituting ability. Proc Natl Acad Sci USA 1997;94: 13648−13653.
31. Pawliuk R, Eaves C, Humphries RK. Evidence of both ontogeny and transplant dose-regulated expansion of hematopoietic stem cells in vivo. Blood 1996;88:2852−2858.
32. Sauvageau G, Thorsteinsdottir U, Eaves CJ, Lawrence HJ, Largman C, Lansdorp PM et al. Overexpression of HOXB4 in hematopoietic cells causes the selective expansion of more primitive populations in vitro and in vivo. Genes Dev 1995;9:1753−1765.
33. Fraser CC, Szilvassy SJ, Eaves CJ, Humphries RK. Proliferation of totipotent hematopoietic stem cells in vitro with retention of long-term competitive in vivo reconstituting ability. Proc Natl Acad Sci USA 1992;89:1968−1972.
34. Audet J, Miller CL, Rose-John S, Eaves CJ, Piret JM. In vitro expansion of in vivo repopulating hematopoietic stem cells from adult mouse bone marrow using HyperIL-6, an engineered hybrid cytokine of human interleukin-6 and its soluble receptor. Exp Hematol 1998;26:700 (Abstract).
35. Dick JE. Normal and leukemic human stem cells assayed in SCID mice. Sem Immunol 1996;8:197−206.
36. Zanjani ED, Flake AW, Rice H, Hedrick M, Tavassoli M. Long-term repopulating ability of xenogeneic transplanted human fetal liver hematopoietic stem cells in sheep. J Clin Invest 1994;93:1051−1055.
37. Cashman J, Bockhold K, Hogge DE, Eaves AC, Eaves CJ. Sustained proliferation, multilineage differentiation and maintenance of primitive human haematopoietic cells in NOD/SCID mice transplanted with human cord blood. Br J Haematol 1997;98:1026−1036.
38. Conneally E, Cashman J, Petzer A, Eaves C. Expansion in vitro of transplantable human cord blood stem cells demonstrated using a quantitative assay of their lympho-myeloid repopulating activity in nonobese diabetic-*scid/scid* mice. Proc Natl Acad Sci USA 1997;94:9836−9841.
39. Wang JCY, Doedens M, Dick JE. Primitive human hematopoietic cells are enriched in cord blood compared with adult bone marrow or mobilized peripheral blood as measured by the

quantitative in vivo SCID-repopulating cell assay. Blood 1997;89:3919–3924.

40. Bhatia M, Bonnet D, Kapp U, Wang JCY, Murdoch B, Dick JE. Quantitative analysis reveals expansion of human hematopoietic repopulating cells after short-term ex vivo culture. J Exp Med 1997;186:619–624.

41. Larochelle A, Vormoor J, Hanenberg H, Wang JCY, Bhatia M, Lapidot T et al. Identification of primitive human hematopoietic cells capable of repopulating NOD/SCID mouse bone marrow: implications for gene therapy. Nature Med 1996;2:1329–1337.

Signal transduction of vascular endothelial growth factor (VEGF) receptors, Flt-1 and KDR/Flk-1

Masabumi Shibuya[1], Tomoko Takahashi[1], Asako Sawano[1], Sachie Hiratsuka[1], Sachiyo Ogawa[1], Naoyuki Yabana[1], Yoshiro Maru[1], Tetsuo Noda[2] and Sachiko Yamaguchi[1]

[1]Department of Genetics, Institute of Medical Science, University of Tokyo, Tokyo; and [2]Department of Cell Biology, Cancer Institute, Tokyo, Japan

Abstract. VEGF has been shown to be crucial for the physiological and most of the pathological angiogeneses. To examine the signal transduction from the two VEGF receptors (VEGFR), Flt-1 and KDR/Flk-1, we established NIH3T3 cell lines overexpressing each of the VEGFRs. We found that most of the mitotic signal was generated from KDR/Flk-1. However, unlike other representative tyrosine kinase receptors, KDR/Flk-1 utilizes the activation of phospholipase-Cγ-protein kinase C pathway for the stimulation of MAP kinase and DNA synthesis, but not or only weakly the activation of Ras or PI3kinase pathway. The idea that most of the positive signals towards endothelial cell proliferation and vascular permeability are mediated by KDR/Flk-1 was confirmed by a novel type VEGF-like molecule (VEGF-E) which only binds and activates KDR/Flk-1. Flt-1 was found to carry a stronger binding affinity to VEGF but a much weaker tyrosine kinase activity. Studies on the *flt-1* tyrosine kinase-deficient mice suggest that the high-affinity ligand-binding domain of the Flt-1 is sufficient for the establishment of physiological angiogenesis, most likely regulating the levels of VEGF to an appropriate range in embryogenesis.

Keywords: angiogenesis, endothelial cell, signal transduction, tyrosine kinase receptor, VEGF.

Introduction

Angiogenesis is an essential process for the establishment of cardiovascular systems and the proper organization of most of the normal tissues of animals [1,2]. The endothelial cell growth factor vascular endothelial growth factor (VEGF), also designated as vascular permeability factor (VPF) has been shown to be involved in a variety of angiogeneses under normal conditions such as embryogenesis, corpus luteum formation and placental growth as well as under pathological conditions such as diabetic retinopathy, rheumatoid arthritis and tumor angiogenesis [3—5] (Fig. 1).

We previously isolated the tyrosine kinase receptor *flt-1* (*fms*-like tyrosine kinase) gene that encodes the 7-Ig domain containing extracellular region and a tyrosine kinase domain carrying an insert of about 70 amino acids (kinase insert) in the middle of this domain [6]. Flt-1 and its related tyrosine kinase

Address for correspondence: Masabumi Shibuya, Department of Genetics, Institute of Medical Science, University of Tokyo, 4-6-1 Shirokanedai, Minato-ku, Tokyo 108-8639, Japan. Tel.: +81-3-5449-5550. Fax: +81-3-5449-5425. E-mail: shibuya@ims.u-tokyo.ac.jp

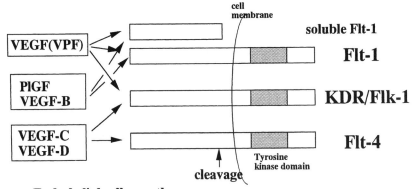

Fig. 1. VEGF and its receptor system. A relationship between VEGF family members and its receptors are indicated by arrows.

KDR/Flk-1 are high-affinity receptors for VEGF [7–9]. Furthermore, a VEGF-related ligand placenta growth factor (PlGF) expressed in placental tissue and a related ligand VEGF-B bind Flt-1 but not KDR/Flk-1 [10–12]. However, the signal transduction and the functions of these VEGF receptors (VEGFRs) are still poorly understood.

In general, VEGFRs are coexpressed in primary endothelial cells [13] and downregulated in most of the established endothelial cell lines. Therefore, first we examined the signaling from each of the VEGFRs overexpressed in NIH3T3 cells, then studied the signal transduction in primary endothelial cells. Further, we utilized a new type of VEGF, originally identified in a parapox viral genome [14], which was found to bind only one of the VEGFRs [15]. The existence of a ligand which stimulates only one of the receptors is a good tool to dissect the signaling generated by VEGF. Finally we examined the importance of the extracellular domain of the Flt-1 by domain-specific gene targeting of mice.

Results and Discussion

Signal transduction of VEGFRs expressed in NIH3T3 cells

^{35}S-methionine-labeled VEGF-binding studies revealed that NIH3T3 cells do not express either VEGFR Flt-1 or KDR/Flk-1 at detectable levels [16]. In addition, NIH3T3 cells are widely used as a good cell culture system to examine the signal transduction and transforming activity of a variety of receptor-type tyrosine kinases such as EGF receptor. Therefore, we constructed an expression vector for Flt-1 or KDR/Flk-1, and introduced them to NIH3T3 cells by DNA transfection method. We established stable cell lines overexpressing Flt-1 (5,000 to 10,000 receptors per cell) or KDR/Flk-1 (10,000 to 20,000 receptors per cell) [11,16].

Using these cell lines we first examined the affinity of the Flt-1 or KDR/Flk-1 to VEGF or PlGF. We found that the Kd value of the association of Flt-1 with VEGF was 2–10 pM, whereas that of PlGF was about 170 pM. Interestingly, the kD value of the association of KDR/Flk-1 with VEGF was about 410 pM, one order weaker than that of Flt-1 to VEGF. KDR/Flk-1 did not show a binding activity to PlGF [11].

In terms of the tyrosine kinase activity, however, we observed an opposite relationship. KDR/Flk-1 showed a relatively strong kinase activity and autophosphorylation on tyrosine, whereas the Flt-1 carried only 10-fold less tyrosine kinase activity [11,16,17]. Therefore, although the two VEGFRs show a quite similar structural relationship to each other, biochemical activities appear to be quite different from each other [18]. Furthermore, stimulation of NIH3T3-Flt-1 with VEGF did not show any detectable growth promoting activity [16]. Stimulation of NIH3T3-KDR/Flk-1 with the same ligand exerted cell proliferation [17], although its activity was weaker than the case of EGFR-overexpressing NIH3T3 cells stimulated with EGF.

These results suggest that the main signal transducer in endothelial cells for cell proliferation is KDR/Flk-1 kinase.

Processing and signal transduction of KDR/Flk-1

Next we examined the processing and the signaling of KDR/Flk-1 in NIH3T3 cells. Pulse-chase experiments using radioisotope-labeled methionine and glycosidase-digestion experiments indicated that about 150 kDa nascent KDR/Flk-1 protein was first synthesized and rapidly glycosylated to an intermediate form of the 210-kDa KDR/Flk-1 protein which is not expressed on the cell surface. After a slow processing with further glycosylation, a mature form of 230-kDa KDR/Flk-1 product was synthesized and expressed on the cell surface as a functional receptor [17] (Fig. 2).

This processing of KDR/Flk-1 in NIH3T3 cells appear to be quite similar to that in the endogenously expressed KDR/Flk-1 in primary endothelial cells, although the sizes of the intermediate and mature forms are slightly different from those expressed in NIH3T3 cells, probably due to a difference in the levels of glycosylation.

An interesting question is how the signal transduction of KDR/Flk-1 is similar to (or different from) that of the representative tyrosine kinase receptor such as EGF receptor or PDGF receptor. KDR/Flk-1 belongs to the FLT(7Ig-TKR)/PDGF receptor (5Ig-TKR) supergene family. KDR/Flk-1 shares a long kinase-

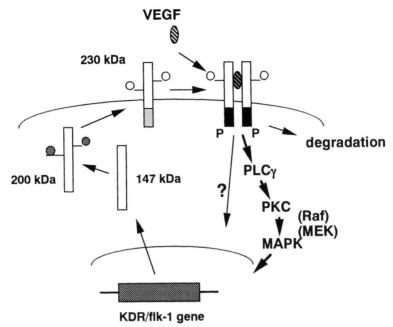

VEGF

230 kDa

200 kDa   **147 kDa**   **?**

P   **P** → **degradation**

PLCγ

PKC   **(Raf)**
          (MEK)
MAPK

KDR/flk-1 gene

Tunicamycin: both 230 kDa and 200 kDa change to 180 kDa
Autophosphorylation: (+) in 230 kDa, but (-) in 200 kDa

Fig. 2. Processing and signal transduction of KDR/Flk-1 expressed in NIH3T3 cells.

insert sequence with PDGF receptor gene family in the tyrosine kinase domain. However, tyrosine-X-X-methionine motifs in the kinase insert region of the PDGF receptor family, which is known to be important for PI3-kinase activation and Ras activation were not conserved in that of KDR/Flk-1.

We found that one of the major target molecules of KDR/Flk-1 tyrosine kinase is PLCγ [18]. Upon stimulation with VEGF, PLCγ was rapidly associated with KDR/Flk-1 and tyrosine-phosphorylated. Further, MAP kinase was activated in response to VEGF stimulation.

To examine the possible involvement of protein kinase C downstream of PLCγ in VEGF-induced MAP kinase activation in NIH3T3 cells, protein kinase C was downregulated with TPA. The TPA-treated NIH3T3-KDR/Flk-1 cells did not show any MAP kinase activation nor stimulation of DNA synthesis, although the tyrosine phosphorylation of KDR/Flk-1 and of PLCγ was not affected. Thus, these results indicate that KDR/Flk-1 preferentially utilizes PLCγ-protein kinase C-MAP kinase pathway for signal transduction [18] (Fig. 2).

Recently we have confirmed that KDR/Flk-1 in primary endothelial cells also utilizes a similar pathway, PLCγ-PKC-Raf-MEK-MAP kinase pathway towards DNA synthesis, and this was activated more efficiently than that in NIH3T3-KDR/Flk-1 cells. The time course of MAP kinase activation in primary endothelial cells after stimulation with VEGF was rapid, peaked at 5—10 min compared with that (peaked at 20 to 30 min) in NIH3T3-KDR/Flk-1 cells.

Furthermore, Wortmannin, a PI3-kinase-specific inhibitor or dominant negative Ras did not block significantly the KDR/Flk-1-mediated MAP kinase activation nor DNA synthesis [19]. Taken together, these results strongly suggest that KDR/Flk-1 is unique among tyrosine kinase receptors, which utilizes PLCγ-PKC-MAP kinase pathway as a major route for the stimulation of DNA synthesis in vascular endothelial cells.

A novel type VEGF-like molecule, VEGF-E/NZ-7 VEGF

In 1994, an open reading frame which encodes a protein distantly related to VEGF was reported in the genome of NZ-2 and NZ-7 strains in the Orf virus of parapox virus, which preferentially infect sheep and goat, occasionally humans [14]. We purified the protein encoded in the genome of NZ-7 strain by Baculovirus expression system, and found that the protein has a homodimer structure, and binds and stimulates the autophosphorylation of only one of the VEGFRs, KDR/Flk-1 [13]. This VEGF-like molecule is quite useful to dissect the functions of two VEGFRs Flt-1 and KDR/Flk-1. We propose a name VEGF-E for this protein because of an alphabetical designation of the VEGF family (Fig. 3).

VEGF carries two major biological functions, growth stimulation of endothelial cells and enhancement of vascular permeability. Using VEGF-E as a tool, we examined whether or not VEGF-E shows these two activities. We found that VEGF-E exerted both activities, growth stimulation of endothelial cells and increase in vascular permeability in Miles assay similar to VEGF, and the levels of these activities associated with VEGF-E were almost the same as those with VEGF-165, the most potent subtype of VEGF. These results are well consistent with the idea that the most of the two biological activities associated with VEGF is mediated by KDR/Flk-1 [13].

Biological analysis of the flt-1 gene in vivo using tyrosine kinase-deficient mutant mice

The *flt-1* null mutant mice were reported to be embryonic lethal due to disorganization of blood vessels and overgrowth of endothelial-like cells in the vessels [20]. Although the tyrosine kinase of Flt-1 could generate PLCγ-mediated MAP kinase activation and p38 MAP kinase activation to some extent, generally these activities associated with Flt-1 are much weaker to those with KDR/Flk-1 [21]. Since the extracellular domain of the Flt-1 bears a tight binding activity with VEGF, we examined whether the tyrosine kinase domain of Flt-1 is dispensable in physiological angiogenesis.

In order to prepare the tyrosine kinase-deficient *flt-1* gene, we first analyzed the *flt-1* genomic structure in mice, and found that the exon 17 encodes the amino-terminal part of the Flt-1 tyrosine kinase domain [22]. The exon 17 was

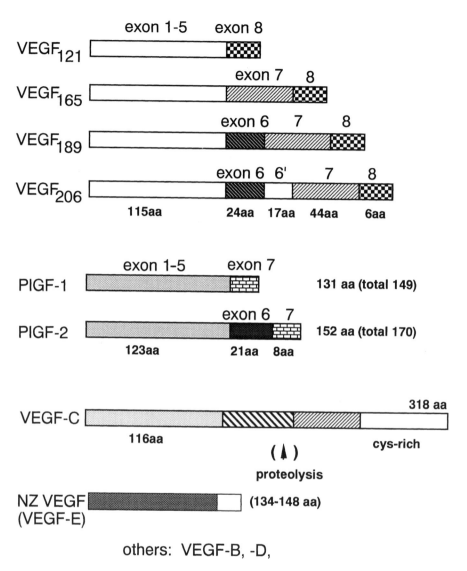

Fig. 3. Structural relationship among the VEGF family. VEGF-E (NZ-7 virus-encoded VEGF-like molecule) shows 23–25% amino acid-identity with VEGF-121, but carries potent mitogenic and vascular permeability activities [15].

replaced by a *neo* gene cassette by homologous recombination, and heterozygous mice were mated to generate homozygous mice [23].

Consistent with our hypothesis that the extracellular domain of Flt-1 is more important than the tyrosine kinase domain bearing a weak enzymatic activity, the extracellular domain of the *flt-1* without the cytoplasmic domain was found to be sufficient for normal embryogenesis and angiogenesis [23]. However, cell

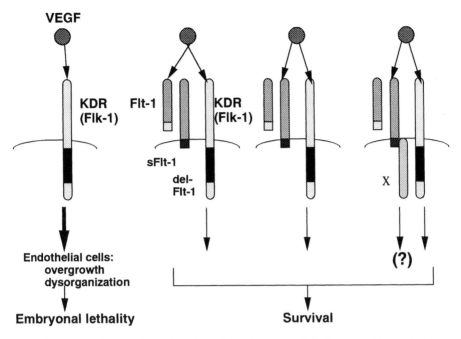

Fig. 4. Several possibilities for explanation of the phenotype of *flt-1* tyrosine kinase-deficient mice. The ligand-binding domain of the *flt-1* is sufficient for normal embryogenesis and angiogenesis in mice. The simplest explanation for it is that the high-affinity binding domain of the Flt-1 traps the low concentration of VEGF to negatively regulate the levels of endogenous VEGF in embryogenesis (middle two cases). However, a minor possibility that the extracellular domain and the transmembrane domain of the Flt-1 associate with other signal transducer(s) and generate a signal for morphogenesis (right).

migration of macrophages induced by VEGF was severely impaired.

In humans, monocytes/macrophages were reported to express only Flt-1 but not KDR/Flk-1 [24,25]. Therefore, our studies on KDR/Flk-1 signaling and the *flt-1* tyrosine kinase-deficient mice strongly suggest the following models:

1) The major positive signal of VEGF towards endothelial cell proliferation and vascular permeability is mediated by KDR/Flk-1.

2) The extracellular domain of the *flt-1* gene is sufficient for angiogenesis in physiological conditions (Fig. 4).

3) The Flt-1 tyrosine kinase is required for VEGF-induced migration of macrophages, thus the *flt-1* is utilized differently in two types of cells, endothelial cells and macrophages.

It is of quite interest to examine whether tyrosine kinase of Flt-1 plays a role in some pathological angiogenesis such as tumor angiogenesis. We are currently investigating these possibilities.

32

References

1. Risau W, Sariola H, Zerwes H-G, Sasse J, Ekblom P, Kemler R, Doetschman T. Vasculogenesis and angiogenesis in embryonic-stem-cell-derived embryoid bodies. Development 1988;102: 471—478.
2. Folkman J. What is the evidence that tumors are angiogenesis dependent? J Natl Cancer Inst 1990;82:4—6.
3. Ferrara N, Houck K, Jakeman L, Leung DW. Molecular and biological properties of the vascular endothelial growth factor family of proteins. Endocrinol Rev 1992;13:18—32.
4. Shibuya M. Role of VEGF-Flt receptor system in normal and tumor angiogenesis. Adv Cancer Res 1995;67:281—316.
5. Mustonen T, Alitalo K. Endothelial receptor tyrosine kinases involved in angiogenesis. J Cell Biol 1995;129:895—898.
6. Shibuya M, Yamaguchi S, Yamane A, Ikeda T, Tojo A, Matsushime H, Sato M. Nucleotide sequence and expression of a novel human receptor-type tyrosine kinase gene (flt) closely related to the fms family. Oncogene 1990;5:519—524.
7. de Vries C, Escobedo JA, Ueno H, Houck K, Ferrara N, Williams LT. The fms-like tyrosine kinase, a receptor for vascular endothelial growth factor. Science 1992;255:989—991.
8. Millauer B, Wizigmann-Voos S, Shuruch H, Martinez R, Moller NPH, Risau W, Ullrich A. High affinity VEGF binding and developmental expression suggest flk-1 as a major regulator of vasculogenesis and angiogenesis. Cell 1993;72:835—846.
9. Terman BI, Dougher-Vermazen M, Carrion ME, Dimitrov D, Armellino DC, Gospodarowicz D, Bohlen P. Identification of the KDR tyrosine kinase as a receptor for vascular endothelial growth factor. Biochem Biophys Res Commun 1992;187:1579—1586.
10. Park JE, Chen HH, Winer J, Houck KA, Ferrara N. Placenta growth factor: potentiation of vascular endothelial growth factor bioactivity, in vitro and in vivo, and high affinity binding to Flt-1 but not to Flk-1/KDR. J Biol Chem 1994;269:25646—25654.
11. Sawano A, Takahashi T, Yamaguchi S, Aonuma T, Shibuya M. Flt-1 but not KDR/Flk-1 tyrosine kinase is a receptor for placenta growth factor (PlGF), which is related to vascular endothelial growth factor (VEGF). Cell Grow Diff 1996;7:213—221.
12. Olofsson B, Korpelainen E, Mandriota S, Pepper MS, Aase K, Kumar V, Gunji Y, Jeltsch MM, Shibuya M, Alitalo K, Eriksson U. VEGF-B binds to VEGFR-1 and regulates plasminogen activator activity in endothelial cells. Proc Natl Acad Sci USA 1998;(In press).
13. Yamane A, Seetharam L, Yamaguchi S, Gotoh N, Takahashi T, Neufeld G, Shibuya M. A new communication system between hepatocytes and sinusoidal endothelial cells in liver through vascular endothelial growth factor and Flt tyrosine kinase receptor family (Flt-1 and KDR/ Flk-1). Oncogene 1994;9:2683—2690.
14. Lyttle DJ, Fraser KM, Fleming SB, Mercer AA, Robinson AJ. Homologs of vascular endothelial growth factor are encoded by the poxvirus orf virus. J Virol 1994;68:84—92.
15. Ogawa S, Oku A, Sawano A, Yamaguchi S, Yazaki Y, Shibuya M. A novel type of vascular endothelial growth factor: VEGF-E (NZ-7 VEGF) preferentially utilizes KDR/Flk-1 receptor and carries a potent mitotic activity without heparin-binding domain. J Biol Chem 1998;273: 31273—31282.
16. Seetharam L, Gotoh N, Maru Y, Neufeld G, Yamaguchi S, Shibuya M. A unique signal transduction from FLT tyrosine kinase, a receptor for vascular endothelial growth factor (VEGF). Oncogene 1995;10:135—147.
17. Takahashi T, Shibuya M. The 230 kDa mature form of KDR/Flk-1 (VEGF receptor-2) activates the PLCg pathway and partially induces mitotic signals in NIH3T3 fibroblasts. Oncogene 1997;14:2079—2089.
18. Shibuya M, Ito N, Claesson-Welsh L. Structure and function of VEGF Receptor-1 and -2. Curr Topic Microbiol Immunol 1999;(In press).
19. Takahashi T, Ueno H, Shibuya M. VEGF activates protein kinase C-dependent, but Ras-

independent Raf-MEK-MAP kinase pathway for DNA synthesis in primary endothelial cells. Oncogene 1998;(In press).

20. Fong G-H, Rossant J, Gertsentein M, Breitman ML. Role of the Flt-1 receptor tyrosine kinase in regulating the assembly of vascular endothelium. Nature 1995;376:66—70.

21. Sawano A, Takahashi T, Yamaguchi S, Shibuya M. The phosphorylated 1169-tyrosine containing region of Flt-1 kinase (VEGFR-1) is a major binding site for PLCγ. Biochem Biophys Res Commun 1997;238:487—491.

22. Kondo K, Hiratsuka S, Subbalakshmi E, Matsushime H, Shibuya M. Genomic organization of the *flt-1* gene encoding for vascular endothelial growth factor (VEGF) receptor-1 suggests an intimate evolutionary relationship between the 7-Ig and the 5-Ig tyrosine kinase receptors. Gene 1998;208:297—305.

23. Hiratsuka S, Minowa O, Kuno J, Noda T, Shibuya M. Flt-1 lacking the tyrosine kinase domain is sufficient for normal development and angiogenesis in mice. Proc Natl Acad Sci USA 1998;95:9349—9354.

24. Barleon B, Sozzani S, Zhou D, Weich HA, Mantovani A, Marme D. Migration of human monocytes in response to vascular endothelial growth factor (VEGF) is mediated via the VEGF receptor flt-1. Blood 1996;87:3336—3343.

25. Clauss M, Weich H, Breier G, Knies U, Rock W, Waltenberger J, Risau W. The vascular endothelial growth factor receptor Flt-1 mediates biological activities. J Biol Chem 1996;271: 17629—17634.

Tissue Engineering for Therapeutic Use 3.
Y. Ikada and T. Okano, editors.

Liver stem-like (oval) cells: isolation, differentiation and application for cell therapy

Toshihiro Sugiyama, Ouki Yasui and Naoyuki Miura
Department of Biochemistry Akita University School of Medicine, Akita, Japan

Abstract. We isolated oval cells from the liver of Long-Evans Cinnamon (LEC) rats by isopyknic centrifugation in a Percoll gradient. The cells were γ-glutamyl transpeptidase positive, α-fetoprotein positive, cytokeratin 18 and 19 positive, but albumin negative in the cells. Among liver-enriched transcription factors, only HNF-3β mRNA was expressed, but HNF-3α, HNF-4, HNF-1α and C/EBP mRNAs were undetectable. When oval cells were transplanted to the liver, they were transformed into hepatocytes. To evaluate albumin biosynthesis, we transplanted oval cells into the liver of Nagase analbuminemic and LEC double mutant rats. The albumin level in the serum of transplanted rats was increased and maintained for up to 10 weeks. These results indicated that the oval cells isolated from LEC rats can differentiate into hepatocytes in vivo and can be used as a source for cell transplantation therapy to the liver.

Keywords: cell transplantation, gene therapy, hepatocyte, LEC rat, oval cell.

Introduction

Gene therapy to transfer novel genes into the liver has potential application in metabolic/genetic diseases and tumors. Several methods and the application of a gene delivery system have been introduced using nonviral or viral transfer. Several cell types have been used as gene delivery vehicles including white blood cells, keratinocytes, fibroblasts, myoblasts, and hepatocytes. Although hepatocytes can be used for the liver, primary culture and genetic engineering of hepatocytes remain serious technical obstacles to the development of ex vivo gene therapy.

When the liver is severely injured, hepatic regeneration may be accomplished not only by hepatocyte proliferation, but also by the emergence of hepatocyte stem/progenitor cells that subsequently differentiate into hepatocytes (Fig. 1). Oval cells which proliferate during chemically induced severe liver injury, are considered to be liver stem/progenitor cells [1]. Some investigators [2—6] have reported that oval cells resemble biliary epithelial cells in morphological and immunohistochemical features, whereas others [7—9] have reported that oval cells are precursors of hepatocytes. Several oval cell lines have been established from carcinogen-treated rats and they possess both hepatocyte and biliary

Address for correspondence: Toshihiro Sugiyama, Department of Biochemistry Akita University School of Medicine, 1-1-1 Hondo, Akita 010-8543, Japan. Tel. and Fax: +81-18-884-6443. E-mail: sugiyama@med.akita-u.ac.jp

Regeneration after injury

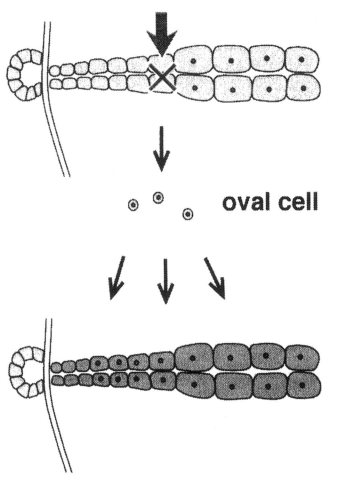

oval cell

Fig. 1. Schematic diagram illustrating the hypothesis that oval cells develop from stem cells located in biliary ductule and that they differentiate into hepatocytes and bile duct.

epithelial cell markers [10—12].

Long-Evans Cinnamon (LEC) rats [13] carry a defect in the Wilson disease gene (*ATP7B* gene) [14] resulting in copper accumulation in the liver. The appearance and characteristics of oval cells in the LEC rat histologically resemble those in rats given hepatic carcinogens.

In this study, we isolated oval cells from LEC rats, characterized them, and then transplanted them into the liver of LEC and Nagase analbuminemic rats (NAR/LEC). We found that our established oval cells were transformed into hepatocytes in vivo and stably produced a liver-specific protein, albumin. As oval cells are cultured and manipulated more easily in vitro, we considered that oval cells could be used as gene delivery vehicles.

Results

Characterization of oval cells

We isolated oval cells from LEC rats and characterized them (Table 1). After plating, the oval cells proliferated and assumed a cobblestone appearance at day 14 (Fig. 2A). Electron microscopic examination of oval cells showed protruding microvilli on the cell surface, a small proportion of cytoplasm, irregularly shaped nuclei with variable amounts of condensed chromatin, small and abundant mitochondria and a few lysosomes (Fig. 2B). These morphological findings indicate that our oval cells had the characteristics of hepatocytes and of biliary epithelial cells.

We examined the markers present in the oval cells. Cytokeratin (CK)18 are expressed in hepatocytes and bile duct epithelium. α-Fetoprotein (AFP) and albumin are characteristic of hepatocytes, while CK19 is a marker of bile duct

Table 1. Phenotypic characteristics of oval cells.

Morphology, ultrastructure and other phenotypic traits	
Diameter of cells	10-15 μm
Nucleus	Oval-shaped, a pale nucleosm
Ultra structure	Protruding microvilli
	Small and abundant mitochondria
	A few lysosomes
Antigenic makers for hepatocytes	
α-Fetoprotein	+
Albumin	−
Cytokeratin 18	+
Antigenic makers for bile ducts	
Cytokeratin 19	+
γ-Glutamyltranspeptidase	+ (gradually decreased)
Gene expressin	
HNF-3β	+
HNF-3α	−
HNF-4	−
HNF-1α	−
C/EBP	−

38

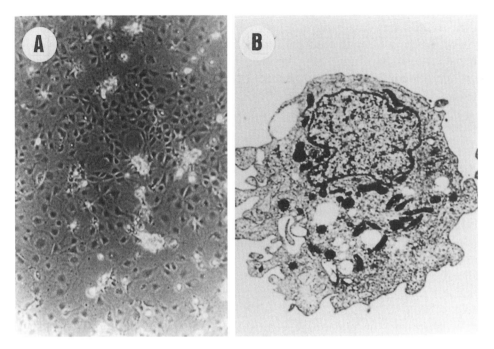

Fig. 2. Phase contrast microscopic (**A**) and electron microscopic (**B**) photographs of oval cells cultured on collagen-coated dishes at day 14. Magnification: **A** × 200; **B** × 4,000.

epithelium. The oval cells at day 14 were immunoreactive against AFP, CK18 and CK19, but were not stained for albumin. During liver development, the liver transcription factors HNF-3β, HNF-3α, HNF-4, HNF-1α and C/EBP are sequentially expressed in cells of hepatocyte-lineage. Among liver-enriched transcription factors, only HNF-3β mRNA was expressed, but HNF-3α HNF-4, HNF-1α and C/EBP mRNAs were undetectable.

The above-mentioned findings suggested that our oval cells have characteristics of hepatocytes and of biliary epithelial cells and that they correspond to immature hepatoblasts.

Transplantation of the oval cells into the liver

As the oval cells had dual characteristics of hepatocytes and biliary epithelial cells, we determined whether or not they can be transformed into hepatocytes or bile duct. We infected the oval cells with a LacZ-transducing retrovirus (SG virus) and transplanted them into the liver of NAR/LEC rats. We examined X-gal stained liver sections from the transplanted NAR/LEC rat and immunohistochemically stained them with anti-albumin antibody. As shown in Fig. 3A,B, NAR/LEC hepatocytes without the transplantation had negligible immunoreactivity for albumin. On the other hand, several areas in the transplanted liver were heavily stained with anti-albumin antibody and the X-gal stained cells were

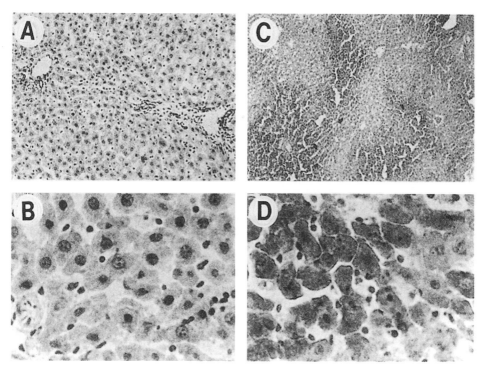

Fig. 3. Transplantation of the LacZ gene carrying oval cells into the liver of NAR/LEC rats. Liver sections from the control (**A** and **B**) and transplanted (**C** and **D**) NAR/LEC rats were stained with X-gal, then immunohistochemically stained with anti-albumin antibody. Magnification: **A** × 200 and **C** × 100; **B** and **D** × 400.

always immunoreactive to albumin (Fig. 3C, D). These results indicated that the transplanted oval cells produced albumin as mature hepatocytes.

Finally, we evaluated the efficiency of the oval cell transplantation by measuring the serum albumin level in the transplanted rats (Fig. 4). In the transplanted rats, serum albumin remained increased for up to 10 weeks, reaching a maximum of 0.3 mg/ml 2 weeks after transplantation then gradually decreased from 10 weeks. At 6 and 10 weeks after transplantation of oval cells, serum albumin decreased slowly. We speculate that transplanted cells may be removed by an immunological response of NAR genetic background.

Discussion

Hepatocytes have been transplanted into the liver. Hepatocyte replacement has been demonstrated by transplanting hepatocytes into transgenic mice with damaged hepatocytes [15]. However, primary culture and genetic manipulation of hepatocytes requires a high degree of skill and long-term culture of hepatocytes is as yet impossible.

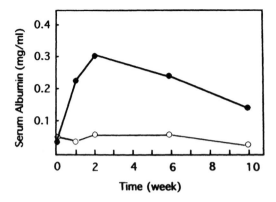

Fig. 4. Elevated albumin levels in the serum of transplanted NAR/LEC rats was maintained for up to 10 weeks. Time course of the serum albumin in the control (open circles) and transplanted (closed circles) NAR/LEC rats. Serum albumin levels were determined by densitometry of the Western blots of serially diluted samples and known amounts of albumin.

Oval cells are considered to be bipotential cells which can be differentiated into hepatocytes and bile ducts [1]. However, when our oval cell line was transplanted into the liver of rats in vivo, it was transformed into hepatocytes, but not into bile ducts.

Gene therapy, the insertion of therapeutic genes into a recipient's cells, holds enormous promise for curing many diseases [16—19]. However, for this therapy to succeed, an efficient method for delivering new active genes into patients must be developed. White blood cells, keratinocytes, fibroblasts, myoblasts and hepatocytes have been used as ex vivo gene delivery systems. Genetically modified myoblasts injected into the muscles can produce high levels of human growth hormone [20,21]. Therefore, immature myoblasts can be used as efficient gene delivery vehicles. Myoblasts could be fused with the recipient's muscle cells and stably maintained to produce new substances.

For ex vivo gene therapy of the liver, oval cells seem better than hepatocytes. In contrast to hepatocytes, oval cells can be cultured for a long time, genetically manipulated in vitro with ease, and transplanted as hepatocytes in vivo. Our gene delivery system, oval cell-mediated hepatocyte transplantation, may be used not only to cure hepatocytes defective in metabolic genes, but also to deliver a wide variety of substances that either act in the blood or are transported by the blood to other tissues. Our oval cells have been cultured for more than 1 year and then can be genetically modified by infection with a recombinant retrovirus and by transfection with plasmid DNAs. Even after a long culture period, they can be transformed into hepatocytes when transplanted into the liver. These findings indicated that our cells can be used as a new gene delivery vehicle.

References

1. Fausto N. Hepatocyte differentiation and progenitor cells. Curr Opin Cell Biol 1990; 2:1036—1042.
2. Dempo K, Chisaka N, Yoshida Y, Kaneko A, Onoe T. Immunofluorescent study on α-fetopro-tein-producing cells in the early stage of 3'-methyl-4-dimethylaminoazobenzene carcinogenesis. Cancer Res 1975;35:1282—1287.
3. Farber E. Similarities in the sequence of early histological changes induced in the liver of the rat by ethionine, 2-acetylaminofluorene, and 3'-methyl-4-dimetylaminoazobenzene. Cancer Res 1956;16:142—155.
4. Grisham JW, Hartroft WS. Morphologic identification by electron microscopy of "oval cells" in experimental hepatic degeneration. Lab Invest 1961;10:317—332.
5. Lenzi R, Liu MH, Tarsetti F, Slott PA, Alpini G, Zhai W-R, Paronetto F, Lenzen R, Tavoloni N. Histogenesis of bile duct-like cells proliferating during ethionine hepatocarcinogenesis. Lab Invest 1992;66:390—402.
6. Sell S, Salman J. Light- and electron-microscopic autoradiographic analysis of proliferating cells during the early stages of chemical hepatocarcinogenesis in the rat induced by feeding N-2-fluorenylacetamide in a choline-deficient diet. Am J Pathol 1984;114:287—300.
7. Evarts RP, Nagy P, Marsden E, Thorgeirsson SS. In situ hybridization studies on expression of albumin and α-fetoprotein during the early stage of neoplastic transformation in rat liver. Cancer Res 1987;47:5469—5475.
8. Evarts RP, Nagy P, Marsden E, Thorgeirsson SS. A precursor-product relationship exists between oval cells and hepatocytes in rat liver. Carcinogenesis 1987;8:1737—1740.
9. Evarts RP, Nagy P, Nakatsukasa H, Marsden E, Thorgeirsson SS. In vivo differentiation of rat liver oval cells into hepatocytes. Cancer Res 1989;49:1541—1547.
10. Pack R, Heck R, Dienes HP, Oesch F, Steinberg P. Isolation, biochemical characterization, long-term culture, and phenotype modulation of oval cells from carcinogen-fed rats. Exp Cell Res 1993;204:198—209.
11. Braun L, Goyette M, Yaswen M, Thompson NL, Fausto N. Growth in culture and tumorigeni-city after transfection with the ras oncogene of liver epithelial cells from carcinogen-treated rats. Cancer Res 1987;47:4116—4124.
12. Tsao MS, Smith JD, Nelson KG, Grisham JW. A diploid epithelial cell line from normal adult rat liver with phenotypic properties of "oval" cells. Exp Cell Res 1984;154:38—52.
13. Yoshida MC, Sasaki M, Masuda R. Origin of the LEC strain with a new mutation causing hereditary hepatitis. In: Mori M, Yoshida MC, Takeichi N, Taniguchi N (eds) The LEC Rat, A New Model for Hepatitis and Liver Cancer. Tokyo: Springer-Verlag, 1991;3—10.
14. Wu J, Forbes JR, Chen HS, Cox DW. The LEC rat has a deletion in the copper transporting ATPase gene homologous to the Wilson disease gene. Nature Genet 1994;7:541—545.
15. Rhim JA, Sandgren EP, Degen JL, Palmiter RD, Brinster RL. Replacement of diseased mouse liver by hepatic cell transplantation. Science 1994;263:1149—1152.
16. Anderson WF. Human gene therapy. Science 1992;256:808.
17. Miller AD. Human gene therapy comes of age. Nature 1992;357:455.
18. Lyerly HK, DiMaio JM. Gene delivery systems in surgery. Arch Surg 1993;128:1197—1206.
19. Mulligan RC. The basic science of gene therapy. Science 1993;260:926—932.
20. Barr E, Leiden JM. Systemic delivery of recombinant proteins by genetically modified myo-blasts. Science 1991;254:1507—1509.
21. Dhawan J, Pan LC, Pavlath GK, Travis MA, Lanctot AM, Blau HM. Systemic delivery of human growth hormone by injection of genetically engineered myoblasts. Science 1991;254: 1509—1512.

Development of a hybrid artificial liver support system and its application to hepatic failure animals

Kohji Nakazawa[1], Hiroyuki Ijima[1], Mitsuru Kaneko[1], Junji Fukuda[1], Tomonobu Gion[2], Mitsuo Shimada[2], Ken Shirabe[2], Kenji Takenaka[2], Keizo Sugimachi[2] and Kazumori Funatsu[1]

Departments of [1]Chemical Systems and Engineering, Graduate School of Engineering, and [2]Surgery II, Faculty of Medicine, Kyushu University, Fukuoka, Japan

Abstract. We found that primary rat, dog and porcine hepatocytes form multicellular aggregates (spheroids) in the pores of polyurethane foam (PUF) as a culture substratum, and that they maintain several liver-specific functions, such as albumin secretion and urea synthesis, for at least 2 weeks in vitro. Using these spheroids in PUF, we developed a hybrid artificial liver support system (HALSS), that induces an extracorporeal circulation system, and a hepatocyte culture module. The geometry of the module was a packed bed whose total volume was filled with a PUF block with many capillaries supplying the medium. About 10% of the hepatocytes in a whole liver were inoculated into the PUF pores and formed spheroids. This HALSS was applied to various animals with hepatic failure. In the experiment with rats induced with D-galactosamine, 80% of rats with hepatic failure recovered by applying HALSS including a module packed with 5.0×10^7 cells (n = 5). Furthermore, in large animal experiments, using dogs (15 kg) and pigs (25 kg) with ischemic liver failure, the blood glucose concentrations were well maintained and the increases in blood ammonia concentration were suppressed by applying the HALSS including the modules packed with 3.0×10^9 hepatocytes for dog and 6.4×10^9 hepatocytes for pig. These results indicate that HALSS seems promising for clinical application.

Keywords: extracorporeal circulation, hepatocyte spheroid culture, polyurethane foam, primary hepatocyte spheroid, tissue engineering.

Introduction

Recently, the development of a hybrid artificial liver support system (HALSS), composed of an extracorporeal circulation system, hepatocytes culture module, etc., has been expected as a real possibility in order to cure and save fulminant hepatic failure patients. For this purpose, HALSS must have selective removal abilities of toxic substances, and synthesis activities of essential metabolites [1]. Many researchers have tried to develop an effective hepatocyte culture module, and some researchers have already applied their modules to patients as a bridge to liver transplantation [2,3].

Address for correspondence: Kazumori Funatsu, Department of Chemical Systems and Engineering, Graduate School of Engineering, Kyushu University, 6-10-1 Hakozaki, Higashi-ku, Fukuoka 812-8581, Japan. Tel.: +81-92-642-3508. Fax: +81-92-642-3513.

For the development of a practical HALSS, it is firstly important to express and maintain high-level functions of hepatocytes for as long as possible. In a mono-layer culture of primary hepatocytes, which is a conventional culture technique, its hepatic functions rapidly decrease within a few days. On the other hand, three-dimensional culture of multicellular aggregates (spheroids) has been reported by many researchers in recent years [4—8], and these spheroids express higher liver-specific functions for a period as long as 2—3 weeks, presumably because spheroids have a tissue-like structure similar to in vivo tissue, thereby maintaining the functions to a certain degree. Consequently the technique of reorganizing from isolated and dispersed cells to a tissue-like structure is very important. Much effort should be made to find new techniques for making a tissue-like structure. The basic functions of a reorganized tissue-like structure are of great importance for the development of HALSS with high performance. Brief explanations of advances in the concept of a tissue-like structure are shown in Fig. 1.

In this article, we describe our original technique of spheroid formation of hepatocytes and the development of HALSS using a spheroid culture and the evaluation of its performance using hepatic failure animals. Furthermore, we ten-tatively estimate the scale of our HALSS for clinical applications.

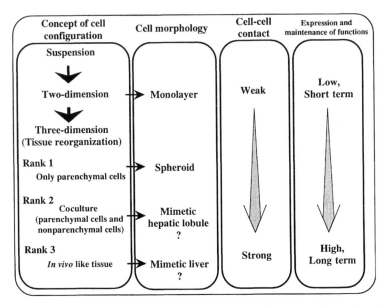

Fig. 1. Advance in the concept of tissue-like structure from two to three dimensions. Because of the advances in culture techniques for hepatocytes from two-dimensional culture, such as the conventional monolayer, to three-dimensional culture, such as multicellular aggregates (spheroids), it is hoped that hepatocytes in tissue-like structure can express and maintain liver-specific functions in the long term. Thus, the advance in tissue engineering may be a key point to develop a high-perfor-mance HALSS.

Three-dimensional culture of various primary hepatocytes using polyurethane foam (PUF) as a culture substratum

As mentioned above, the first technical points for the development of HALSS are the expression and maintenance of functions in hepatocytes at a high level in vitro: three-dimensional culture is an effective method at present.

We used a polyether rigid-type PUF (INOAC, Japan) composed of isocyanate and polyol as a culture substratum. It has a sponge-like macroporous structure with each pore made up of smooth thin films and thick skeletons (average pore size: about 300 µm, porosity: 98.8%), as shown in Fig. 2.

Primary hepatocytes of various animals such as rat, dog, and pig, which were inoculated into PUF, formed multicellular aggregates (spheroids) with a range of 100–150 µm in diameter within 24 h (Fig. 3). The spheroids formed spontaneously, and attached to the bottom surface of PUF pores. This means that immobilized spheroids can be used as a biocatalyst such as immobilized cells, immobilized enzymes, etc.

Although it is well known that the liver-specific functions of hepatocytes in monolayer culture rapidly decrease in a few days, the functions such as albumin secretion, urea synthesis and drug metabolism can be maintained in the hepatocyte spheroid culture for at least 2 weeks [9–12].

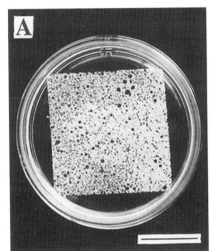
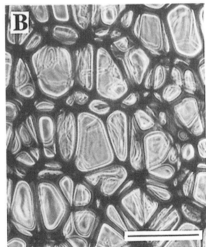

Fig. 2. PUF as a cell culture substratum. **A:** Appearance of PUF plate (bar: 2 cm). **B:** Microphotograph of PUF (bar: 500 µm). PUF consists of isocyanate and polyol as main materials and has a sponge-like macroporous structure with each pore made up of smooth thin films and thick skeletons (average pore size: about 300 µm, porosity: 98.8%). PUF pores are partially opened and connected with one another. Its surface is weakly hydrophobic and has characteristics of peptide bond-like structure. The hepatocytes attach to the surface and form spheroids.

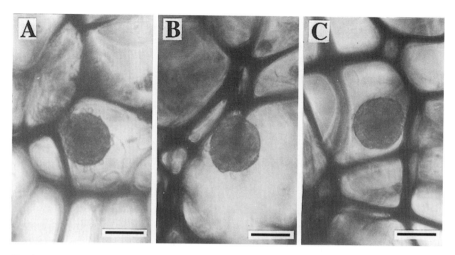

Fig. 3. Spheroid formation of primary hepatocytes of various animals. **A:** Rat hepatocytes. **B:** Dog hepatocytes. **C:** Porcine hepatocytes. Bar is 100 μm. Primary rat, dog, and porcine hepatocytes, which were inoculated into PUF, formed multicellular aggregates (spheroids) with a range in diameter of 100–150 μm within 24 h. The spheroid formation of hepatocytes is spontaneous, and they consist of about 200 hepatocytes. The process of spheroid formation in PUF pore is as follows: 1) attaching and spreading of hepatocytes in the internal surface of PUF pore, 2) peeling off and assembling of cells, and 3) forming spheroids. Furthermore, the functions of hepatocyte spheroid culture, such as albumin secretion, urea synthesis and drug metabolism, can be maintained stably for at least 2 weeks in vitro.

HALSS

Hepatocyte culture module using PUF packed bed with multicapillary

HALSS must be designed compactly in order to be installed at the bedside of a patient. Consequently, since the hepatocyte culture module must be made compactly, it must achieve high cell density culture and sufficient mass transfer for the supply of oxygen and nutrients and the removal of waste metabolites.

PUF can achieve a high cell density of more than 1.0×10^7 cells/cm^3 of PUF. After experimental and theoretical studies, we developed a hepatocyte culture module of multicapillary PUF packed bed [13], as shown in Fig. 4. This module using a cylindrical PUF block has many capillaries for the flow of the culture medium or plasma, which have a diameter of 1.5 mm drilled in a triangular arrangement of 3.0 mm. The culture medium or plasma flows into these capillaries and directly penetrates the connected pores of PUF that are partially opened and connected with each other. Therefore, a good mass transfer can be achieved between the medium and the hepatocyte spheroids in the pores of PUF. In addition, scale up of the module can be carried out comparatively easily by multiplying, for example, the ratio of scaled up total cells to the present number of total cells, to the present volume, because the situations concerning the

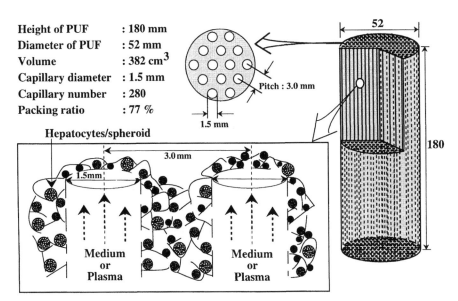

Height of PUF	: 180 mm
Diameter of PUF	: 52 mm
Volume	: 382 cm^3
Capillary diameter	: 1.5 mm
Capillary number	: 280
Packing ratio	: 77 %

Fig. 4. Details of a hepatocyte culture module for dog experiments. This figure shows a hepatocyte culture module of multicapillary PUF packed bed for dog experiments. This module uses cylindrical PUF block (diameter: 52 mm, height: 180 mm) and has many capillaries as flow channels for the culture medium or plasma. These capillaries with a diameter of 1.5 mm are drilled in a triangular arrangement of 3.0 mm. Hepatocyte spheroids form in the macroporous structure between each of the capillaries. The culture medium or plasma flow into these capillaries can directly penetrate the connected pores of PUF, because PUF pores are partially opened and connected with one another. Therefore, a good mass transfer can be achieved between the medium and the hepatocyte spheroids in the pores of PUF.

structure of PUF, medium flow, mass transfer, etc., can be considered to be almost identical in all localization, including three capillaries arranged in a triangle, even if it scaled up or down.

We made hepatocyte culture modules packed with 10% of the hepatocytes of a whole liver for the application to rats, dogs, and pigs with hepatic failure, and we evaluated the scale-up criteria.

Extracorporeal circulation system including hepatocyte culture module

As a prototype of HALSS, we made an extracorporeal circulation system, which was a modified blood purification device. The system is composed of a blood circulation loop on the side of the living body and a plasma circulation loop on the side of the hepatocyte culture module, and both loops are connected by a plasma separator (Fig. 5) [14]. Toxic substances contained in the blood permeate the plasma separator to the hepatocyte culture module side. In the hepatocyte culture module side, the hepatocytes detoxify the toxic substances in the plasma and produce glucose and other compounds, and the treated plasma circulates back to the living body side.

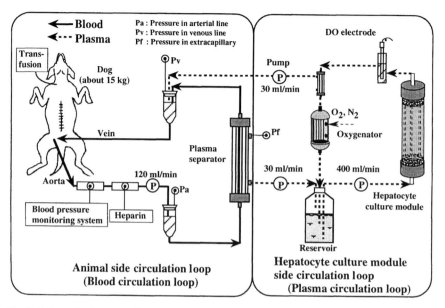

Fig. 5. Schematic diagram of HALSS for dog experiments. This system is composed of the blood circulation loop on animal side and the plasma circulation loop on the hepatocyte culture module side, and both loops are connected with a plasma separator. Toxic substances in the blood permeate the plasma separator to the hepatocyte culture module side. On the hepatocyte culture module side, hepatocytes clean the plasma and produce glucose and other compounds. The treated plasma circulates back to the animal side. (Reproduced, with permission, from [14].)

Performance of HALSS

All animal experiments were carried out under the Guidelines for Animal Experiments in Kyushu University in the Animal Center in Kyushu University.

Rat experiments

We made the hepatocyte culture module (volume: 5.6 cm^3, height: 5.0 cm, diameter: 1.2 cm) packed with 5.0×10^7 rat hepatocytes (0.5 g) for the rat experiments according to our above-mentioned design scheme. We applied it to hepatic failure rats (body weight: about 200 g) induced by intraperitoneal injection of D-galactosamine. The performance of our HALSS was evaluated by comparison with the results of the control experiment whose system was the same except that it lacked hepatocytes.

In the control experiment, the blood ammonia concentration rapidly increased during 3 h of circulation, and all of the rats died (n = 3). On the other hand, in the HALSS experiment, the increase in blood ammonia was suppressed to less than the hepatic coma level (200 μg/dl) during and after the circulation, and 80% of hepatic failure rats were recovered by applying the HALSS (n = 5) [11,12,15].

Dog experiments

For the hepatocyte culture module in the dog experiments, we made a larger module (volume: 382 cm^3, height: 18.0 cm, diameter: 5.2 cm) packed with 3.0×10^9 dog hepatocytes (30 g) corresponding to about 10% of the hepatocytes of a whole liver in the dog (body weight: about 15 kg), and applied it to a dog (15 kg) with warm ischemic liver failure caused by a portocaval shunt. Because this was a severe acute liver failure model, we mainly measured the biochemical compositions in the blood during the initial periods of HALSS treatment. Furthermore, only the electrolyte solution was supplied during the circulation period in order to evaluate the definite effect of the hepatocyte culture module alone. The performance of HALSS was evaluated by comparison with the results of the control experiment which was the same system except that it contained no hepatocytes.

The performances of HALSS treatment in the dog experiments are summarized in Table 1. The blood ammonia concentration dramatically increased from the initial stage in the control experiment. This drastic increase in ammonia indicates induction of severe acute liver failure. The increase in the blood ammonia concentration was suppressed by HALSS. The blood glucose concentration rapidly decreased in the control experiment, but increased initially in the HALSS experiment, where it was maintained better than in the control experiment [14].

Table 1. Performance of HALSS treatment in dog and pig experiments. This table shows the time course of blood ammonia (index of detoxification) and glucose (index of metabolism) in the warm ischemic hepatic failure dogs and pigs in the treatment with our HALSS. Because this model is a severe acute liver failure model, it is important to establish the biochemical compositions in the blood during the initial periods of HALSS treatment. By applying HALSS, the increases in the blood ammonia concentrations were suppressed, and the blood glucose concentrations were well maintained. These results show the availability of our HALSS treatment.

	Ammonia and glucose concentration	Circulation time (h)			Normal level of concentration
		0	1	4	
Dog	Blood ammonia (N-μg/dl)				
	HALSS (n = 2)	96 ± 0	225 ± 23	206 ± 37	30–85
	Control (n = 2)	192 ± 48	313 ± 88	459 ± 89	
	Blood glucose (mg/dl)				
	HALSS (n = 2)	208 ± 13	349 ± 42	228 ± 84	70–120
	Control (n = 2)	215 ± 46	189 ± 6	65 ± 23	
Pig	Blood ammonia (N-μg/dl)				
	HALSS (n = 1)	119	149	174	30–85
	Control (n = 1)	137	195	—	
	Blood glucose (mg/dl)				
	HALSS (n = 1)	229	239	255	70–120
	Control (n = 1)	203	114	—	

Pig experiments

In the pig experiments, we used two large modules of the same dimensions that were made for the dog experiments. The total cell number was 6.4×10^9 porcine hepatocytes (64 g), corresponding to about 10% of the hepatocytes of a whole liver in a pig (body weight: about 25 kg). The performance of HALSS was evaluated using a pig (25 kg) with warm ischemic liver failure caused by a portocaval shunt. The conditions and methods of performance evaluation were the same as in the dog experiments.

The performances of HALSS treatment in the pig experiments are summarized in Table 1. The survival times of the hepatic failure pigs in the control and HALSS experiment were 2.6 and 10.5 h, respectively. The blood glucose concentration decreased dramatically in the control experiment, but was well maintained in the HALSS experiment. The blood ammonia concentration was more suppressed in the HALSS experiment than in the control experiment.

Scale-up criteria of HALSS for humans

From the results of the series of experiments using small to large animals, it may be concluded that HALSS, which is designed and operated in our scheme, can be applicable to various animals for cure of hepatic failure. We must check and improve the details, but almost the same design scheme of HALSS can be used in application to humans. Thus, we tried to scale up HALSS for 70-kg patients, for example, from the results of the animal experiments. In one trial, we used basic parameters such as body weight, blood volume in the body, and liver weight as listed in Table 2, to estimate the necessary weight of hepatocytes packed in the module, the volume in the module, and other quantities. In this case, the volume of the hepatocyte culture module for clinical application became 1,960 or 5,880 cm^3, which is 2.8 times that of the pig.

Conclusions

In this article, we briefly show an outline of the development of our HALSS. Presently, we are carrying out pig experiments as preclinical applications. In progressing to clinical application, the following problems must be solved: 1) mass preparation of porcine hepatocytes, 2) fabrication of a large and sophisticated hepatocyte culture module and extracorporeal circulation system, 3) suppression of immunological reaction, 4) prevention of pathogen infection from HALSS to human, etc. In the near future, by solving these problems, we hope that practical HALSS can be a cure and save fulminant hepatic failure patients.

Table 2. Scale-up criteria of HALSS for various animals. This table shows scale-up criteria of HALSS packed with about 10 and 30% of the hepatocytes in a whole liver for various animals. The size of human HALSS can be estimated from the results of the animal experiments and the basic parameters of the corresponding living body. Using the values for rat, the hepatocyte culture module for dog and pig were 60 and 100 times, respectively, scaled up compared to the size of a rat. The hepatocyte weights are from 196 to 588 g and the module volume are from 1,960 to 5,880 cm^3 for a human.

Animal		Human	Pig	Dog	Rat
Basic parameter of a living body	Body weight (kg)	70	25	15	0.25
	(ratio)	(280)	(100)	(60)	(1)
	Blood volume (ml)	5600	1375	1290	16
	(ratio)	(350)	(86)	(81)	(1)
	Liver weight (g)	1500	500	300	7
	(ratio)	(214)	(71)	(43)	(1)
Hepatocyte culture module in HALSS	Cell weight (g)				
	10% of the hepatocytes of a whole liver[a]	196	70	42	0.7
	30% of the hepatocytes of a whole liver[b]	588	210	126	2.1
	(Ratio)	(280)	(100)	(60)	(1)
	Volume (cm^3)				
	10% of the hepatocytes of a whole liver[a]	1960	700	420	7
	30% of the hepatocytes of a whole liver[b]	5880	2100	1260	21
	(Ratio)	(280)	(100)	(60)	(1)

Numbers in parentheses are the ratios to the value for rat. [a]Minimum quantity for liver support by medical doctors; [b]hint from necessary quantity for living liver transplantation.

Acknowledgements

This study was supported by Grant-in-Aids for Developmental Scientific Research (A) (2): 08555206 and (A) (2): 10305062 from the Ministry of Education, Science, Sports and Culture, and by Kyushu University Interdisciplinary Programs in Education and Projects in Research Development of Type B.

References

1. Funatsu K, Kawakubo Y, Ijima H, Matsushita T. High density culture of animal cells and its application to hybrid artificial liver support systems in Japan. Artif Organs Today 1994;3: 253–274.
2. Gerlach JC. Development of a hybrid liver support system: a review. Int J Art Organ 1996;19: 645–654.
3. Chen SC, Hewitt WR, Watanabe FD, Eguchi S, Kahaku E, Middleton Y, Rozga J, Demetriou AA. Clinical experience with a porcine hepatocyte-based liver support system. Int J Art Organ 1996;19:664–669.
4. Koide N, Sakaguchi K, Koide Y, Asano K, Kawaguchi M, Matsushota H, Takenami T, Shinji T,

Mori M, Tsuji T. Formation of multicellular spheroids composed of adult rat hepatocytes in dishes with positively charged surfaces and under other nonadherent environments. Exp Cell Res 1990;186:227—235.

5. Tong JG, Lagausie PD, Furlan V, Cresteil T, Bernard O, Alvarez F. Long-term culture of adult rat hepatocyte spheroids. Exp Cell Res 1992;200:326—332.

6. Takezawa T, Yamazaki M, Mori Y, Yonaha T, Yoshizato K. Morphological and immunocyto-chemical characterization of a heterospheroid composed of fibroblasts and hepatocytes. J Cell Science 1992;101:495—501.

7. Lazar A, Mann HJ, Remmel RP, Shatford RA, Cerra FB, Hu WS. Extended liver-specific functions of porcine hepatocyte spheroids entrapped in collagen gel. In Vitro Cell Dev Biol 1995; 31:340—346.

8. Sakai Y, Naruse K, Nagashima I, Muto T, Suzuki M. Large-scale preparation and function of porcine hepatocyte spheroids. Int J Art Organ 1996;19:294—301.

9. Matsushita T, Ijima H, Koide N, Funatsu K. High albumin production by multicellular spheroid of adult rat hepatocytes formed in the pores of polyurethane foam. Appl Microbiol Biotechnol 1991;36:324—326.

10. Matsushita T, Ijima H, Funatsu K. Development of a hybrid type artificial liver utilizing three dimensional culture of adult hepatocytes. Jpn J Art Organ 1992;21:1050—1054.

11. Ijima H, Matsushita T, Nakazawa K, Fujii Y, Funatsu K. Hepatocyte spheroids in polyurethane foams: functional analysis and application for a hybrid artificial liver. Tis Eng 1998;4:213—226.

12. Ijima H, Nakazawa K, Mizumoto H, Matsushita T, Funatsu K. Formation of a spherical multi-cellular aggregate (spheroid) of animal cells in the pores of polyurethane foam as a cell culture substratum and its application to a hybrid artificial liver. J Biomater Sci 1998;9:765—778.

13. Ijima H, Matsushita T, Funatsu K. Development of a hybrid artificial liver using multicapillary PUF/spheroid packed-bed. Jpn J Art Organ 1994;23:463—468.

14. Matsushita T, Koyama S, Ijima H, Nakazawa K, Gion T, Shirabe K, Shimada M, Takenaka K, Sugimachi K, Funatsu K. Application of hybrid artificial liver using PUF/hepatocytes-spheroid packed-bed module to warm ischemic liver failure dog. Jpn J Art Organ 1997;26:455—459.

15. Matsushita T, Ijima H, Wada S, Funatsu K. Estimation of the performance of a PUF/spheroid packed-bed type artificial liver by using an extracorporeal circulation with hepatic failure rats. Jpn J Art Organ 1995;24:815—820.

JSPS TISSUE ENGINEERING PROJECT: BIOPROCESS ENGINEERING OF
FUNCTIONAL REGENERATION OF CULTURED ANIMAL CELLS

Improvement of the performance of a packed-bed bioartificial liver

Norio Ohshima, Kennichi Yanagi, Hirotoshi Miyoshi and Pei Kan
Department of Biomedical Engineering, Institute of Basic Medical Sciences, University of Tsukuba, Tsukuba Science City, Japan

Abstract. To construct a bioartificial liver, a culture technique to enable high-density and large-scale culture of hepatocytes so as to maintain viability and metabolic functions of cultured hepatocytes comparable to those in vivo is required. In order to achieve this goal the following attempts are under progress:

Hepatocytes cultured in a packed-bed reactor (cell density: 8.6×10^6 cells/cm^3-PVF), using reticulated polyvinyl formal (PVF) resin as a cell-supporting material, under high dissolved oxygen concentration ranging from 260 to 460 μM showed 30% higher ammonium metabolic activity and 85% higher albumin secretion activity compared with those of the monolayer culture.

Two kinds of nonparenchymal cells, i.e., nonparenchymal liver cells obtained from rats (NPC) and porcine aortic endothelial cells (EC), were adopted for coculture experiments using plate-shaped PVF resin. When hepatocytes were cocultured with NPC, no significant effects were observed. On the other hand, the hepatocytes cocultured with EC maintained longer albumin secretion than the hepatocytes cultured solely.

Effects of shear stress ranging from 1 to 14 dyne/cm^2 on the coculture system of hepatocyte/NPC were investigated. Hepatocyte aggregates were formed and their aggregation was facilitated by exposure to shear flow. Significant increases in metabolic functions were evoked by shear flow.

Keywords: coculture, hepatocyte, nonparenchymal liver cell, polyvinyl formal resin, shear stress.

Introduction

A bioartificial liver that uses heterologous living hepatocytes cultured in a mass transfer device is needed to assist metabolic functions of patients suffering from severe liver insufficiency such as fulminant hepatitis. One of the difficulties in developing a bioartificial liver is that the hepatocytes show less ability to proliferate under usual culture conditions, and very easily lose their viability [1,2]. To develop a clinically applicable bioartificial liver, a technique to achieve efficient large-scale and high-density culture of hepatocytes is indispensable. We have shown that a new type of a packed-bed reactor utilizing cultured hepatocyte on some porous polymer, a reticulated polyvinyl formal (PVF) resin (Kanebo Kasei

Address for correspondence: Prof Norio Ohshima, Institute of Basic Medical Sciences, University of Tsukuba, 1-1-1 Tennoudai, Tsukuba Science City, Ibaraki-ken 305-8575, Japan. Tel.: +81-298-53-3084. Fax: +81-298-53-3039.

Co., Osaka, Japan), as a support material resulted in a maximum cell density of hepatocytes as high as 1×10^7 cells/cm^3-PVF [2—4]. The resin has a characteristic three-dimensionally reticulated structure with continuous interconnecting pores within its matrix and has substantially high porosity of about 90%. Such a porous and reticulated structure seems suited for high-density immobilization of hepatocytes. Moreover, hepatocytes immobilized at a high density in the PVF resin showed metabolic activities comparable with those in the conventionally used monolayer culture [2—7]. To improve performances of the reactor, a few of following approaches are attempted in our group, and substantial progresses have thus far been made.

Improvement of metabolic functions of a packed-bed reactor by controlling oxygen concentrations in the culture medium

Maintaining a higher level of metabolic functions of cultured hepatocytes for a relatively long term is an essential requirement to realize the bioartificial liver. Oxygen is presumed to be a key factor in maintaining metabolic functions of the cultured hepatocytes, since the liver is known to consume a large quantity of oxygen in vivo. We have recently succeeded in improving the metabolic performances of the hepatocytes cultured in the packed-bed reactor by controlling oxygen concentrations of the culture medium [8].

Materials and Methods

A packed-bed reactor (20 mm i.d.) loaded with 250 PVF resin cubes ($2 \times 2 \times 2$ mm, mean pore size: 250 μm) was used [5,8]. Hepatocytes ($8.5—9.9 \times 10^7$ cells) obtained from male Wistar rats were inoculated into the reactor. Culture medium was perfused from the reservoir by use of a roller pump into the reactor through an oxygenator at a flow rate of 17 ml/min. In all culture experiments, serum-containing medium composed of Williams' medium E, hormones (dexamethasone, 0.1 μM; insulin, 0.1 μM; aprotinin, 5,000 kIU/l), antibiotics (penicillin G, 20,000 IU/l; streptomycin, 20 mg/l; amphotericin B, 50 μg/l) and 10% fetal bovine serum was used. Dissolved oxygen tension of the medium was measured at the outlet of the reactor by use of a dissolved oxygen electrode. Monolayer cultures of hepatocytes using 35-mm Petri dishes (Falcon 1008, Becton Dickinson Labware, Lincoln Park, New Jersey, USA) coated with type-I collagen were simultaneously performed as controls. Dissolved oxygen tension was controlled by changing the gas mixing ratio of air:O_2:CO_2 or air:N_2:CO_2 supplied into the oxygenator (O_2 gas ratio ranged between 10 and 50%). All culture experiments were performed for up to 4 days, and metabolic performances of the cultured hepatocytes were estimated in terms of ammonium metabolism, albumin secretion, alanine aminotransferase (ALT) release, DNA content and ultrastructural morphology as described before [4,6,8].

Results and Discussion

Eight runs of the perfusion culture experiments using the packed-bed reactor and monolayer cultures as controls were performed for 4 days. Dissolved oxygen concentration measured by means of the oxygen electrode placed at the outlet of the reactor, ranged from 60 to 460 μM in each run. Immobilized cell density in the PVF resin varied from 3.8 to 9.7×10^6 cells/cm^3-PVF after completion of the perfusion experiments. As shown in Table 1, immobilized cell density showed significantly higher values at higher dissolved oxygen concentrations.

As measures of hepatocytes viability, ALT activity in the medium was measured. It was found that ALT activity observed in the earlier stage of culture (day 1) showed considerably higher values in all culture conditions, and that those values decreased markedly afterwards (data not shown). No significant differences in ALT activity were observed among culture conditions.

To estimate the ammonium metabolic performance of a single hepatocyte, values of the ammonium metabolic rate constant (km) were calculated as described before [7]. Under high oxygen concentrations of 260–460 μM, hepatocytes cultured in the packed-bed reactor showed 30% higher ammonium metabolic activity than those of the monolayer culture on day 2 ($p < 0.05$). On the other hand, oxygen concentrations below 100 μM impaired ammonium metabolic activity of the cultured hepatocytes.

As shown in Table 1, albumin secretion rate of hepatocytes cultured in the reactor showed almost the same tendency as ammonium metabolic performance. Albumin secretion rate of hepatocytes cultured in the packed-bed reactor showed 85% higher value than those of the monolayer hepatocytes cultured on dishes on day 1 ($p < 0.05$).

Figure 1 shows representative scanning electron photomicrographs (SEM) of hepatocytes immobilized in PVF resins cultured at higher oxygen concentrations. As reported previously [4–6], hepatocytes did not exhibit flattened configurations as found in the monolayer culture, instead, individual cells showed a spheri-

Table 1. Metabolic functions of hepatocytes cultured in differed conditions.

Parameter (unit)	Monolayer	Packed-bed reactor		
		Low oxygen	Normal oxygen	High oxygen
Oxygen concentration (μM)	200[a]	60–80	140–180	260–460
Cell density (10^6 cells/cm^3)	0.58 ± 0.10[b]	5.6 ± 2.6	6.5 ± 1.8	8.6 ± 1.4
ALT (mIU/h/10^6 cells)	4.8 ± 3.0	5.6 ± 1.2	5.4 ± 0.46	4.0 ± 1.1
Km (10^{-9} l/cell/h)	1.2 ± 0.27	0.68 ± 0.89	1.5 ± 0.62	1.7 ± 0.13
Albumin (μg/h/10^6 cells)	3.6 ± 2.1	1.8 ± 0.95	5.6 ± 2.5	6.7 ± 1.7
Number of runs (–)	8	2	2	4

Values are mean ± SD. ALT, alanine aminotransferase; Km, ammonium metabolic rate constant, [a]Calculated from oxygen tension of atmosphere; [b]10^6 cells/dish.

Fig. 1. Scanning electron photomicrograph of hepatocytes immobilized in PVF resin cultured at oxygen concentration of 300 μM. Bar: 50 μm.

cal shape. No cell aggregates such as spheroids were seen in the present experiments.

To realize a clinically applicable bioartificial liver, minimum requirements such as to attain a high-density hepatocyte culture and to maintain its metabolic function high enough as comparable to those in vivo should be met. Oxygen is considered to be one of the factors of prime importance in affecting metabolic functions of the cultured hepatocytes, because the liver is a highly oxygen-consuming organ. In the present study, modification of culture conditions so as to attain higher oxygen concentration in the medium seemed to provide a higher driving force for oxygen transfer inside the PVF resin cube, and consequently higher immobilized cell density in the PVF resins and enhanced metabolic performances of hepatocytes were achieved. In addition, our recent findings exhibited that a relatively high level of oxygen concentration (300—400 μM) in the medium improved metabolic performances of the cultured hepatocytes even in a monolayer.

In conclusion, the effects of oxygen concentration of the culture medium on the metabolic performances of hepatocytes cultured in the packed-bed reactor were clarified. Hepatocytes cultured under culture conditions of higher oxygen concentrations from 260 to 460 μM (cell density: 8.6×10^6 cells/cm^3-PVF) showed 30% higher ammonium metabolic activity and 85% higher albumin secretion activity in the earlier stage of the culture (up to 2 days) compared with the monolayer culture. While, low oxygen concentrations below 100 μM impaired activities of cultured hepatocytes.

Use of coculture system of hepatocytes and nonparenchymal cells

As an attempt to maintain viability and metabolic functions of hepatocytes, we performed coculture experiments by modifying the same culture method utilizing PVF resin mentioned above. In the previous study, some kinds of nonparenchymal cells, such as liver epithelial cell line [9], sinusoidal endothelial cells [10], and nonparenchymal liver cells [11], have been adopted for the cocultured cells with hepatocytes. In these studies, it was found that the cocultured hepatocytes maintained ability of albumin secretion long enough, and that their DNA synthesis was markedly stimulated. In the present study, therefore, we attempted to use two kinds of nonparenchymal cells, i.e., nonparenchymal liver cells (NPC) and endothelial cells (EC) for coculture experiments, and performed long-term coculture experiments for up to 20 days utilizing PVF resin as a substrate material.

Materials and Methods

Hepatocytes were isolated from male Wistar rats weighing 150—250 g by the collagenase perfusion technique [4]. In the coculture experiments, nonparenchymal liver cells (NPC) were obtained from Wistar rats, and aortic endothelial cells (EC) were from porcine aortae.

Similarly to the above-mentioned experiments, highly porous and reticulated polyvinyl formal (PVF) resin was used as a substratum of hepatocytes [2—8]. Based on the results of previous studies of the authors, the resins that have a mean pore size of 250 μm were used. These resins were cut into forms of plate (20 × 20 × 2 mm), and used for supporting materials of hepatocytes.

Stationary culture experiments utilizing the PVF resin as well as monolayer culture experiments on Petri dishes chosen as controls were simultaneously performed. In the stationary culture experiments, the plate-shaped PVF resins were autoclaved, placed in the 35-mm Petri dishes, and washed with culture medium. Hepatocytes (2.5×10^6 cells) with or without nonparenchymal cells were inoculated by irrigating 1 ml of the culture medium from the top layer of PVF resins, and (NPC, 2.5×10^6 cells; EC, 5×10^5 or 2.5×10^6 cells) immediately followed by addition of 1.5 ml of fresh medium. Thereafter, the hepatocytes immobilized into the PVF plates were cultured in 2.5-ml medium. In the monolayer culture experiments, 1×10^6 hepatocytes with or without nonparenchymal cells (NPC, 1×10^6 cells; EC, 2×10^5 or 1×10^6 cells) were inoculated onto 35-mm Petri dishes coated with type-I collagen, and cultured in 2-ml medium. Culture medium was changed everyday. On day 1, day 3 and every 3 days thereafter, 1-mM ammonium chloride containing medium was supplied to determine the activities of ammonium metabolism and urea synthesis of the cultured hepatocytes [7]. The spent media were sampled everyday, and were subjected to measurement of albumin concentrations.

Ammonium, urea-N and albumin concentration in the media were measured by the routine method as described previously [4,6].

Results and Discussion

To estimate the activities of cultured hepatocytes, ammonium metabolic and urea-N synthetic activities in all culture experiments are measured (data not shown). Both in the monolayer and stationary cultures, ammonium metabolic activities of sole hepatocyte culture were considerably reduced at an early stage of the culture experiments maintained for up to about 10 days, and thereafter, relatively stable activities were maintained. These activities of cocultured hepatocytes with NPC showed similar results to those of the control hepatocyte culture, and showed decrease with the elapse of culture period. When hepatocytes were cocultured with EC, ammonium metabolic activities were decreased faster than the control culture, and this tendency was particularly prominent in the cocultured hepatocytes with a large amount of EC. These facts probably seemed due to the ammonium production by EC. Concerning the results of urea-N synthetic activities which are final metabolites of ammonium, a similar trend with the results of ammonium metabolic activities was obtained. Thus, no significant beneficial effects of the coculture system with NPC or EC on ammonium metabolic and urea-N synthetic activities of hepatocytes were observed.

The results of albumin secretion activities obtained from all culture experiments are summarized in Table 2. In contrast to the results of ammonium metabolism, some beneficial effects of coculture on albumin secretion ability of the cultured hepatocytes were observed. In the control hepatocyte cultures, albumin secretion rates in the monolayer culture experiments were precipitously decreased in an earlier culture period up to day 6. These metabolic rates then showed a gradual decrease with the culture period, and at the end of the culture experiments, only about 6% of the albumin secretion activities measured on day 1 were preserved. In contrast, in the stationary cultures using PVF plates, albumin secretion rates decreased more rapidly than those in the monolayer cultures until day 4. Thereafter, however, albumin secretion activities were satisfactorily well maintained, and roughly 60% of the albumin secretion rates measured on day 1 were preserved up to the end of the culture experiments (20 days).

When hepatocytes were cocultured with NPC, addition of NPC did not affect ability of albumin secretion of hepatocytes both in the monolayer and stationary cultures. On the other hand, when hepatocytes were cocultured with EC, prolonged albumin secretion activity was observed, especially in the stationary culture using PVF. In this case, albumin secretion activities decreased rapidly with time at an early stage. These activities, however, turned to increase thereafter, and even at the end of culture, these increases still prevailed.

For these several years, we have applied reticulated PVF resin as a supporting material of the hepatocyte culture [1—8]. In our previous studies, we have shown that a packed-bed reactor using PVF resins has a high ability to attain high-density culture of hepatocytes. We also have shown that metabolic performances of the packed-bed reactor, in particular in terms of albumin secretion ability, were satisfactorily maintained over 1 week [6—8]. In the present study, we

Table 2. Time course changes in albumin secretion rates.

Culture period (day)	Hep (μg/h/dish)	Hep+NPC	Hep+EC (5:1)	Hep+EC (1:1)
Monolayer cultures				
1	3.365	2.946	3.140	2.546
2	2.865	2.429	3.294	2.233
4	1.573	1.615	2.582	2.801
6	0.891	0.940	1.685	0.593
8	0.825	0.835	1.335	0.180
10	0.676	0.690	0.519	0.093
12	0.540	0.573	0.065	0.132
14	0.449	0.502	0.052	0.125
16	0.316	0.398	0.050	0.173
18	0.260	0.375	0.068	0.069
20	0.193	0.394	0.076	0.026
Stationary cultures				
1	1.823	2.109	1.186	1.373
2	1.406	1.282	0.763	0.682
4	1.161	1.194	0.426	0.276
6	1.224	1.261	0.731	0.332
8	1.224	1.297	1.264	0.811
10	1.113	1.082	1.506	1.252
12	1.063	0.956	1.235	1.283
14	1.038	1.022	1.508	1.274
16	1.019	1.029	1.399	1.289
18	0.978	1.003	1.849	1.313
20	1.062	1.030	2.069	1.301

Hep: hepatocytes; NPC: nonparenchymal liver cells; EC: endothelial cells.

attempted to investigate the effects of coculture system of hepatocytes with non-parenchymal cells for a longer period of about 20 days by modifying our culture methods using PVF. From these experiments, it was shown that albumin secretion activities in the stationary cultures were proven to be sufficiently well preserved throughout the culture periods. Moreover, when hepatocytes were cocultured with EC, increased activity of albumin secretion of the hepatocytes were also observed in the stationary culture.

From the SEM observations, it was confirmed that the immobilized hepatocytes revealed a spherical shape individually in the stationary cultures [7]. In the coculture experiments with EC, an abundant amount of extracellular matrixes produced by EC were also observed [12]. Thus, it was strongly suggested that these morphological changes in hepatocyte shape and the existence of extracellular matrixes improved albumin secretion activities of the cultured hepatocytes. Thus, a further study to elucidate the effects of cell shape and extracellular matrixes on liver specific functions is essentially needed.

In conclusion, the advantages of the hepatocyte culture technique utilizing PVF resin over the conventional dish culture and those in addition of EC in maintaining some representative metabolic function specific to hepatocytes were clarified.

Effects of shear stress on metabolic functions of the coculture system

Perfusion of the culture medium has been adopted in bioartificial liver devises [1,2,13] to facilitate effective mass exchange of toxic and/or nutrient substances between the medium and cultured hepatocytes. In such situations, shear flow inevitably imposed on hepatocytes by the perfusion of medium was thought to affect the metabolic functions and morphology of hepatocytes. Optimum culture conditions with an appropriate flow rate acting on hepatocytes were thought to be necessary for a prototype bioartificial liver. Alternatively, the hepatocyte system cocultured with nonparenchymal cells has been advocated to maintain and promote hepatocyte activity [12,14]. Certain nonparenchymal cells encompassing hepatic sinusoid are likely to be signal transmitter responding to blood flow changes. Moreover, blood flow may infiltrate sinusoidal fenestra, thereby directly stimulating hepatocytes. Shear stress is thus presumed to play a potential role in inducing the coculture system. In this study, we examined the ammonium metabolism, ureagenesis and morphology of the hepatocytes under variable culture conditions [15]. The presence of shear stress potentially altered hepatocyte activity in many respects.

Materials and Methods

The hepatocytes (5×10^4 cell/cm^2) and nonparenchymal cells (2×10^5 cell/cm^2) were seeded on a type-I collagen-coated rectangular glass plate. Shear flow was applied using a parallel plate flow chamber as described before [15]. The flow rate was adjusted to attain the shear stress of 4.7 dyne/cm^2. The control cultures were carried out simultaneously on a 35 mm Petri dish coated with type-I collagen. Culture medium with 1 mM ammonium chloride added was changed every day, and ammonium concentration in the medium was measured periodically.

Results and Discussion

In this series of experiments, perfusion was instituted after stationary culture. Results of these experiments are shown in Fig. 2. In runs shown in Fig. 2A, stationary culture was maintained for 1 day, and in Fig. 2B for 8 days. When shear flow was applied, metabolic activity was lower than those during stationary culture over the first 3 days of the culture. A rise in ammonium removal rate observed after day 4 seemed to reflect profound effects of shear flow on metabolic functions in the heterotypic (coculture) systems.

A comparison of urea synthesis capabilities measured between different spans

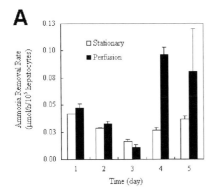 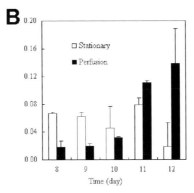

Fig. 2. Ammonium removal by the heterotypic cultures with perfusion starting on day 1 (**A**) and day 8 (**B**), compared with stationary culture. Values are mean ± SD.

of stationary culture performed in advance of perfusion culture was carried out. The stationary culture before the perfusion provided a discernible improvement in terms of urea synthesis activity. An increase in urea synthesis was notable on day 12. In contrast, there was no apparent enhancement in perfusion compared with the stationary system when the term of preperfusion stationary culture was short (1 day).

At the beginning of all culture experiments, hepatocytes were found to form a monolayer of flattened cells. Most of the hepatocytes in the heterotypic stationary cultures remained flattened and surrounded by numerous nonparenchymal cells. On the contrary, the cultures exposed to shear flow showed a distinct cell morphology. A large number of separate aggregates consisting of hepatocytes and nonparenchymal cells were observed over the glass plate. No flattened cells were found in the perfusion system. In addition, hepatocytes were cuboidal in shape within aggregates.

It is inferred that hepatocytes could be influenced explicitly by the blood flow or by induced sinusoidal cells under the conditions adopted. In this study, both hepatocyte and nonparenchymal cells responded to shear flow applied. In view of the decrease in ammonium removal rate found in the present experiments, shear stress is supposed to inhibit the function of hepatocyte in the earlier culture stages. Reorganization and restoration processes are generally expected to take place in the long term. Such a lag time is estimated at about 4 days in the current experiment. After that period, the hepatocyte activity recovered and was enhanced by exposure to shear stress. It is speculated that reorganization of the heterotypic cell system had proceeded, and that the stimulation of shear flow on hepatocyte then due presumably to the cell-cell interactions or secretion of extracellular matrix by the nonparenchymal cells followed.

It was found that the perfusion experiments with variable terms of the preperfusion stationary culture had different outcomes. The perfusion system instituted on day 8 showed a preferential response to shear flow rather than day 1. Accordingly, the change in urea synthesis caused by shear flow was not parallel to that of ammonium removal. An increase in ammonium removal activity in the perfusion systems seems attributable to the action of glutamine cycle, rather than the urea synthesis cycle in hepatocytes.

When perfusion was used, aggregates were observed all over the flow chamber plate. In addition to the potent hepatocyte-nonparenchymal cell interactions, shear flow may alter the level of cell-substratum interaction strength as well as the induced secretion of extracellular matrix by nonparenchymal cells. As a result, shear flow might improve the motility of hepatocytes on the collagen-coated plate, thereby creating aggregates.

In this study, retardation of ammonium removal and urea synthesis was demonstrated particularly in the early period of perfusion. Exposure of hepatocytes to relatively high shear stress is generally considered harmful; however, it is probable that properly reorganized aggregates of hepatocytes cocultured with nonparenchymal cells function well in response to shear flow. Several factors such as cell-cell interactions, hormones and other soluble substances, and extracellular matrix are expected to promote liver-specific functions. Similarly, shear stress and hydrostatic pressure are referred to as latent physical factors. They could participate in the regulation of liver function by way of activation of sinusoidal cells that subsequently are associated with the signal transmission to the hepatocytes.

The shear stress in the normal hepatic sinusoids is estimated to be as high as $20-40$ dyne/cm^2. It is thus suggested that shear flow definitely plays an important role in controlling the liver-specific functions, in view of the results obtained in this study. Therefore, optimization of shear flow and operation conditions are expected to advance the application of bioartificial liver device.

Acknowledgements

The authors thank Ms Tomoko Yoshida, Ms Noriko Sugae and Ms Momoyo Sato for their technical assistance. This research was supported by a grant for "Research for the Future" program (JSPS-RFTF 96I00202) from the Japan Society for the Promotion Science (JSPS).

References

1. Ohshima N. Tissue engineering aspects of the development of bioartificial livers. J Chin Inst Chem Engrs 1997;28:441–453.
2. Ohshima N, Yanagi, K, Miyoshi H. Packed-bed type reactor to attain high density culture of hepatocytes for use as a bioartificial liver. Artif Organs 1997;21:1169–1176.
3. Yanagi K, Mizuno S, Ohshima N. A high density culture of hepatocytes using a reticulated polyvinyl formal resin. ASAIO T 1990;36:M727–M729.
4. Yanagi K, Miyoshi, H, Fukuda H, Ohshima N. A packed-bed reactor utilizing porous resin enables high density culture of hepatocytes. Appl Microbiol Biotechnol 1992;37:316–320.
5. Miyoshi H, Yanagi K, Fukuda H, Ohshima N. Long-term continuous culture of hepatocytes in a

packed-bed reactor utilizing porous resin. Biotechnol Bioeng 1994;43:635–644.

6. Miyoshi H, Yanagi K, Fukuda H, Ohshima N. Long-term performance of albumin secretion of hepatocytes cultured in a packed-bed reactor utilizing porous resin. Artif Organs 1996;20: 803–807.

7. Miyoshi H, Ookawa K, Ohshima N. Hepatocyte culture utilizing porous polyvinyl formal resin maintains long-term stable albumin secretion activity. J Biomater Sci Polymer Edn 1998;9: 227–237.

8. Yanagi K, Miyoshi, H, Ohshima N. Improvement of metabolic performance of hepatocytes cultured in vitro in a packed-bed reactor for use as a bioartificial liver. ASAIO J 1998;44: M436–M440.

9. Guguen-Guillouzo C, Clèment B, Baffet G, Beaumont C, Morel-Chany E, Glaise D, Guillouzo A. Maintenance and reversibility of active albumin secretion by adult rat hepatocytes cocultured with another liver epithelial cell type. Exp Cell Res 1983;143:47–54.

10. Morin O, Normand C. Long-term maintenance of hepatocyte functional activity in co-culture: requirements for sinusoidal endothelial cells and dexamethasone. J Cell Physiol 1986;129: 103–110.

11. Shimaoka S, Nakamura T, Ichihara A. Stimulation of growth of primary cultured adult rat hepatocytes without growth factors by coculture with nonparenchymal liver cells. Exp Cell Res 1987;172:228–242.

12. Miyoshi H, Yanagi K, Ookawa K, Ohshima N. Coculture of hepatocytes using polyvinyl formal resin with two kinds of nonparenchymal cells. Jpn J Artif Organs 1994;23:479–484.

13. Gerlach JC, Schnoy N, Encke J, Smith MD, Muller C, Neuhaus P. Improved hepatocyte in vitro maintenance in a culture model with woven multicompartment capillary systems. Electron microscopy studies. Hepatology 1995;22:546–552.

14. LeCluyse EL, Vullock PL, Parkinson A. Strategies for restoration and maintenance of normal hepatic structure and function in long-term cultures of rat hepatocyte. Adv Drug Del Rev 1996;22:133–186.

15. Kan P, Miyoshi K, Yanagi K, Ohshima N. Effects of shear stress on metabolic function of the coculture system of hepatocyte/nonparenchymal cells for a bioartificial liver. ASAIO J 1998;44: M441–M444.

Tissue Engineering for Therapeutic Use 3.
Y. Ikada and T. Okano, editors.

Repair of adult rat corticospinal tract by transplants of olfactory ensheathing cells

Geoffrey Raisman

The Norman and Sadie Lee Research Centre, National Institute for Medical Research, Medical Research Council, London, UK

Abstract. Injection of a suspension of cultured olfactory ensheathing cells into a unilateral lesion of the upper cervical corticospinal tract in adult rats induces regeneration of cut axons across the lesion and into the distal tract. Whether transplanted at the time of injury or 5 weeks later, the animals recover the ability to acquire a skilled directed paw reaching task on the operated side.

Keywords: axon, regeneration, spinal cord.

Introduction

Repair of tissue falls into two distinct categories. The category requiring less radical change is one in which the tissue can repair itself, such as skin or blood, and here major therapeutic benefit can be obtained by interventions to strengthen the tissues innate powers of repair.

Tissues such as cardiac muscle, kidney, cartilage, and — eminently — the nervous system, present a more difficult problem, since here the tissue has no intrinsic power of self-renewal, and so there is no natural reservoir of stem cells from which loss of tissue can be made up. Improvement depends on the body optimising the surviving tissue (e.g., cardiac hypertrophy, or sprouting in the central nervous system after injury, or in the peripheral nervous system after cell loss, as in the amyotrophic lateral sclerosis (ALS)), and up till now treatment has largely consisted of reducing the load on remaining tissue (such as joints or heart).

In principle, this situation is one which suggests the search for stem cells to add to the surviving tissue and supplement its lost functions (as in Parkinson's or Huntington's diseases). The nervous system, however, presents a unique type of injury — axotomy. The peculiar situation of axotomy arises because of the anatomical arrangement of nerve cells. The nervous system functions as an interactive network of immense complexity. Individual nerve cells constitute the nodes of this network, but in order to make contact with nerve cells at a distance, they produce long processes called axons. The bundling of axons into large fibre tracts recruiting and distributing axons across long distances results in a peculiar type

Address for correspondence: G. Raisman, The Norman and Sadie Lee Research Centre, National Institute for Medical Research, Medical Research Council, The Ridgeway, Mill Hill, London NW7 1AA, UK. Tel.: +44-181-913-8555. Fax: +44-181-913-8587. E-mail: graisma@nimr.mrc.ac.uk

of vulnerability.

One example of this is the lenticulo-striate artery, which lies close to the fibre bundles of the internal capsule. Bleeding from this artery results in immediate and catastrophic disconnection of the cortex from the lower motor control system in the spinal cord. Another major point of vulnerability, which is the subject of our own work, is the spinal cord. The spinal cord is encased in a flexible articulated column of vertebrae, which allows both for protection and for movement. Given the slow pace of natural evolution, however, the vertebral column was unable to adapt to the rate at which scientific and social invention has escalated the violence to which it is subjected.

Axotomy

The neck is the most vulnerable part of the vertebral column. Injuries here threaten the entire complement of bundles of millions of fibres connecting the head and the body. Total destruction results in instant, complete, and irrevocable loss of voluntary movement, inability to breathe, complete loss of bodily sensation, and loss of sexual function and control of bowel and bladder movements. An Ancient Egyptian medical papyrus describing war injuries shows that this type of injury has been recognised by the medical profession for at least 5,000 years. However, until recently, those who suffered from this type of injury did not survive. We are now able to maintain their bodily functions and indeed give them a life expectancy in the normal range, but are unable as yet to offer any method for repairing the original nervous connections whose loss has resulted in their condition. Thus, although life persists, the main symptoms remain unrelieved.

The injury of axotomy severs an axon, leaving a proximal stump attached to the cell body, and a disconnected distal stump. This not only prevents any function (i.e., conduction of impulses), but since the cytoplasm is dependent on the cell body for its maintenance and renewal, the distal stump breaks up, degenerates and the debris is removed by tissue phagocytes.

The nerve cell body reacts in various ways to the loss of its axon. In some cases, the cells survive and the cut ends of the axons form sprouts which explore the wound site, but do not advance through the tissue. In other cases the cells die, either at once, as a result of the ionic fluxes and other physiological disturbances of the injury, or later as a result of lack of growth factors normally obtained by retrograde flow of material taken up in the target area by specialised terminal structures at the tip of the axon. Preventing, or minimising this cell death, for example by introducing exogenous growth factors or modifying the inflammatory response, is one of the first lines of therapeutic response to nerve tract injury.

Once the survival of the axotomised cells has been secured, however, the next stage in repair is to induce the cut axons to regenerate back to their original sites, or to other sites where they can re-establish functionally valuable connections.

Models of repair

There are two "models" for repair of cut axons in the central nervous system. The first is the situation during development, when the axons normally grow to reach their targets. An important goal of current repair strategies is the attempt to induce changes in the pattern of gene expression which will revert the nerve cells to the prior developmental state in which they were originally able to grow axons.

The second model is that of the peripheral nervous system. Here, the proximal ends of cut axons are able to sprout, and provided they have suitable access to the pathways enabling them to elongate as far as their distant targets (such as muscles or sense organs) they are able to re-establish functionally effective connections.

The peripheral nerve "model" has been investigated for over 100 years. In a series of definitive studies Aguayo and colleagues [1] showed that pieces of peripheral nerve grafted into the cut optic nerve (the optic nerve being part of the central nervous system) were able to induce a vigorous growth response in the cut optic nerve fibres and guide the elongation of the sprouts to their original targets in the optic tectum, where they could re-establish anatomical and functional contacts.

While demonstrating the theoretical feasibility of peripherally induced central regeneration, in practice these observations have not yet led to any effective human therapeutic intervention, the numbers of regenerating fibres are small, and it is technically difficult, and in many situations impossible, to introduce grafted peripheral nerve bridges into damaged areas of the central nervous system.

It had been found that for peripheral nerve grafts to succeed in the optic system it is necessary for the grafts to contain living cells [2]. The characteristic cell of the peripheral nervous system is the Schwann cell. To attempt to advance the idea of obtaining some reparative influence from peripheral nerves, a number of groups (e.g., [3]) have cultured Schwann cells and introduced them into central nervous system injuries.

Transplantation into the spinal cord

Our own findings with injection of cultured Schwann cells into the adult central nervous system is that they survive, and indeed migrate and become incorporated into the host tract structure, where they are able to re-myelinate host central axons [4]. Moreover, they have a highly stimulatory effect on the sprouting of the central stump of cut adult corticospinal axons [5]. Disappointingly, these sprouts ramify only locally, forming a dense neuroma, but showing little indication of advancing along their original pathways in the distal part of the denervated host corticospinal tract, as would be needed if they are to reach their original destinations and re-establish functional contacts there.

We therefore sought a source of cells which would combine the regeneration-stimulating properties of Schwann cells with the ability to induce growing axons to leave the wound site, re-enter the central nervous system, and grow back along their original pathways. For this purpose we examined the olfactory system.

The olfactory mucosa contains nerve cells which are exposed to the air stream, where they are subject to continuous physical and chemical damage and infection. Some years ago, it was demonstrated that the olfactory neurons are continually replaced throughout adult life from a stem cell population [6]. This enables the function of olfaction to be maintained despite the continuous attrition of the sensory mucosa, and indeed the rate of cell production is increased following damage [7].

In 1985 [8] I demonstrated the presence of a specialised type of glial cell [9] which chaperones the entry of the olfactory nerve fibres into the olfactory bulb. At their entry into the central nervous system, the terminals of the olfactory nerve fibres are ensheathed by these specialised glia, which accompany them all the way to their first synapses with the arborisations of the primary dendrites of the mitral and tufted cells in the olfactory glomeruli. Doucette [10] has developed a method for culture of these olfactory ensheathing cells, and demonstrated their ability of myelinate axons in culture [11]. In this property they resemble Schwann cells, both in the structure and molecular composition of the myelin they produce, although in olfactory ensheathing cells in situ this myelinating potential is normally latent, as the olfactory nerves are totally unmyelinated in the normal adult.

Injection of cultured olfactory ensheathing cells has been demonstrated to remyelinate demyelinating lesions of the spinal cord [12] and to stimulate growth of dorsal root axons into the spinal cord [13].

We therefore studied the transplantation of cultured olfactory ensheathing cells into unilateral lesions of the upper cervical corticospinal tract in adult rats [14]. We found that the cells survive transplantation, and migrate faster and further than Schwann cells in comparable situations. Most strikingly, they induce rapid, elongative growth of the cut central ends of the adult host corticospinal axons. Quite unlike Schwann cells grafts, which induce the formation of masses of entangled, varicose sprouts, confined locally, the olfactory ensheathing cell grafts induce long slender growing points (with no expanded growth cones), and the unbranched axons pass directly across the lesion area and immediately re-enter the distal part of the host corticospinal tract. Their ultimate distribution is currently under investigation.

From about 3 weeks after transplantation, the olfactory ensheathing cells form peripheral, Schwann-cell-type myelin around the regenerating axons as they pass through the lesion site. Once they enter the distal host corticospinal tract, however, the regenerating axons become remyelinated by central-type oligodendrocytic myelin. The effect of the graft, therefore, is to introduce a "patch" of peripheral type myelin on axons which are myelinated by oligodendrocytic myelin both proximally (in the undamaged part of the corticospinal system) and distally [15].

Functionally, the small, highly localised unilateral upper cervical corticospinal tract lesions have virtually no detectable effect on the health of the animals, or on their motor or sensory performance. To obtain a measure of the functional effectiveness of the repair, therefore, it was necessary to find a defect attributable to the lesion. After considerable investigation we found that, although paw function was apparently normal in all routine activities, the rats were unable to learn to use the paw of the operated side to learn a skilled reaching task in which they were required to direct their paw through a slit in the cage to retrieve a small food pellet.

Histological examination of a number of animals confirmed that this task required complete destruction of the corticospinal tract on that side. Sparing of as few as 1–2% of the fibres spared the function (and this is also significant in terms of the numbers of fibres which would be needed to be repaired to restore the function). Animals with grafts leading to a complete bridging of the lesion showed return of function, and this return could be produced even if the transplantation was delayed for 5 weeks after making the lesion.

Forward look

Whether the regenerating fibres make connections with the original target areas, and in what numbers are matters still under investigation. Conceivably, smaller than normal projections to areas which could ultimately link to the motor mechanism may be enough for the animal to relearn this complex sensori-motor guided behaviour.

We do not know to what extent olfactory ensheathing cells will demonstrate similar reparative properties in other parts of the spinal cord or brain, nor whether it can be transferred to clinical situations. The type of repair we are proposing is one in which tissue is transferred from one part of the body to another. In some ways it is perhaps closest in its approach to skin grafting. It avoids the need for donors, or embryonic tissue, or genetic engineering of transplanted cells.

The observations so far are limited, but encourage the hope that we may be opening the way to one day treating a condition which is still as untreatable today as it was 5,000 years ago when the unnamed military surgeon in Thebes noted the devastating effect of a copper Canaanite arrow.

References

1. Vidal-Sanz M, Bray GM, Villegas-Pérez MP, Thanos S, Aguayo AJ. Axonal regeneration and synapse formation in the superior colliculus by retinal ganglion cells in the adult rat. J Neurosci 1987;7:2894–2909.
2. Berry M, Hall SF, Rees L, Gregson N, Sievers J. Response of axons and glia at the site of anastomosis between the optic nerve and cellular or acellular sciatic nerve grafts. J Neurocytol 1988;17:727–744.
3. Xu XM, Guénard V, Kleitman N, Bunge MB. Axonal regeneration into Schwann cell-seeded guidance channels grafted into transected adult rat spinal cord. J Comp Neurol 1995;351:

145—160.

4. Li Y, Raisman G. Integration of transplanted cultured Schwann cells into the long myelinated fibre tracts of the adult spinal cord. Exp Neurol 1997;145:397—411.

5. Li Y, Raisman G. Schwann cells induce sprouting in motor and sensory axons in the adult rat spinal cord. J Neurosci 1994;14:4050—4063.

6. Graziadei PPC, Montigraziadei GA. Neurogenesis and neuron regeneration in the olfactory system of mammals. I. Morphological aspects of differentiation and structural organization of the olfactory sensory neurons. J Neurocytol 1979;8:1—18.

7. Graziadei PPC, Karlan MS, Montigraziadei GA, Bernstein JJ. Neurogenesis of sensory neurons in the primate olfactory system after section of the fila olfactoria. Brain Res 1980;186:289—300.

8. Raisman G. Specialized neuroglial arrangement may explain the capacity of vomeronasal axons to reinnervate central neurons. Neuroscience 1985;14:237—254.

9. Blanes T. Sobre algunes puntos dudosos de la estructura del bulbo olfactorio. Rev Trim Micro 1898;3:99—127.

10. Doucette JR. PNS-CNS transition zone of the first cranial nerve. J Comp Neurol 1991;312: 451—466.

11. Devon R, Doucette R. Olfactory ensheathing cells myelinate dorsal root ganglion neurites. Brain Res 1992;589:175—179.

12. Franklin RJM, Gilson JM, Franceschini IA, Barnett SC. Schwann cell-like myelination following transplantation of an olfactory bulb-ensheathing cell line into areas of demyelination in the adult CNS. Glia 1996;17:217—224.

13. Ramón-Cueto A, Nieto-Sampedro M. Regeneration into the spinal cord of transected dorsal root axons is promoted by ensheathing glia transplants. Exp Neurol 1994;127:232—244.

14. Li Y, Field PM, Raisman G. Repair of adult rat corticospinal tract by transplants of olfactory ensheathing cells. Science 1997;277:2000—2002.

15. Li Y, Field PM, Raisman G. Regeneration of adult rat corticospinal axons induced by transplanted olfactory ensheathing cells. J Neurosci 1998;18:10514—10524.

JSPS TISSUE ENGINEERING PROJECT: TISSUE ENGINEERING FOR SOFT TISSUES

Regeneration of the peripheral nervous system by artificial nerve conduit

K. Suzuki[1], K. Ohnishi[3], T. Kiyotani[3], G. Lee[3], A.K. Kitahara[1], Y. Suzuki[1], K. Tomihata[3], M. Teramachi[3], Y. Takimoto[3], T. Nakamura[3], K. Endo[2], Y. Nishimura[1], Y. Shimizu[3] and Y. Ikada[3]

Departments of [1]Plastic and Reconstructive Surgery, and [2]Physiology, Faculty of Medicine, and [3]Department of Artificial Organs, Institute for Frontier Medical Sciences, Kyoto University, Kyoto, Japan

Abstract. We demonstrated artificial nerve guidance tubes for short- and long-distance gaps in the case of rat and cat peripheral nerve reconstruction. The gelatin (10 mm), collagen (5 mm) and polyglycolic acid (PGA)-collagen (25 mm) tubes enhanced nerve regeneration and restored parts of function in motor performance. Furthermore, we would like to emphasize the importance of universal ways of evaluation for the novel artificial nerve conduits when earlier clinical application is needed.

Keywords: collagen tube, functional recovery, gelatin tube, long-distance gaps, polyglycolic acid (PGA)-collagen tube, sciatic nerve.

Background

Millions of patients are annually treated in vain for severe traumatic damage as well as progressive degenerative diseases in the central and peripheral nervous systems. Although the central nervous system (CNS) exhibits limited healing and regeneration under almost all conditions, it is generally accepted that the peripheral nervous system (PNS) will regenerate with medical support in addition to its natural capability of heeling. Recent microsurgical techniques and instrumentation have contributed much to the promotion of neural repair. In the 1970s and 1980s, polymer science provided synthetic channels for experimental nerve repair in place of biological materials. From the 1980s to the present, progress in biotechnology and molecular biology provides more knowledge and strategies for enhancing peripheral nerve regeneration [1,2]. We previously developed gelatin and collagen nerve guidance channels and obtained considerable electrophysiological and histological returns in the rat sciatic and cat facial nerve [3,4]. In addition, we established the cat sciatic nerve model in the evaluation of a novel nerve guidance channel and succeeded in bridging a 25-mm gap of the sciatic nerve through the use of a polyglycolic acid (PGA)-collagen nerve conduit from the point of motor performance [5–8]. The PGA-collagen tube is a bio-

Address for correspondence: Katsuaki Endo, Department of Physiology, Faculty of Medicine, Kyoto University, Kyoto 606-8501, Japan. Tel.: +81-75-753-4356. Fax: +81-75-753-4349.
E-mail: endo@med.kyoto-u.ac.jp

material channel made with a combination of reabsorbable polymer, PGA and extracted collagen. In this report, we demonstrate peripheral nerve regeneration across short- and long-distance gaps by our artificial nerve conduits in the rat and cat and discuss their possibilities of clinical application.

Materials and Methods

Preparation of the implanted tubes

Gelatin tube
A silicone tube with a diameter of 1.5 mm was coated on its outer surface with 20% gelatin solution and removed after the gelatin dried. Thus, we made the original form of a gelatin tube. The tube was cross-linked by thermal dehydration at 150°C for 24 h and sterilized using 70% alcohol before surgery [3].

Collagen tube
This guide tube was manufactured from collagen obtained from porcine skin (Nippon Ham, Japan). This collagen was coated over a Teflon tube that was used only as a template, with an external diameter of 2 mm. Then, the coated collagen was dried at room temperature. This step was repeated 20 times. After completion, the tube was subjected to thermal dehydration at 105°C for 24 h. Finally, the Teflon tube was removed and the collagen tube was sterilized using ethylene oxide gas. The manufactured tube measured 2 mm in inner diameter and about 10 mm in length [4].

PGA-collagen tube
We designed a nerve tube of cylindrically woven PGA mesh coated with collagen extracted from porcine skin. The tube was 5 mm in inner diameter and 25 mm in length. The collagen was cross-linked by thermal dehydration at 150°C for 24 h. The tube was sterilized by ethylene oxide gas before surgery [7,8].

Animals and surgery

Rats weighing 200–250 g or cats weighing 3–5 kg were used, respectively. All animals were housed and treated according to the Animal Treatment Guidance of Kyoto University. Under deep anesthesia of sodium pentobarbital (50 mg/kg, ip), the sciatic or facial nerve on the left side was exposed and an approximately 5- to 25-mm segment of the nerve was cut. In rats, a 10-mm gelatin tube was ligated to the distal and proximal nerve stumps with epineural 10.0 nylon threads. In cats, a collagen or PGA-collagen tube filled with the mixture of nerve growth factor (NGF, 100 µg/tube), basic fibroblast growth factor (bFGF, 10 µg/tube) and Matrigel was invaginated into the distal and proximal nerve stumps with epineural 5.0 nylon threads. The right sciatic or facial nerve was left intact and used as controls [3,4,7,8].

Electrophysiological study

Compound muscle action potentials (CMAPs) in the anterior tibialis or gastro-cnemius muscle of the operated side, with electrical stimulation of the sciatic nerve proximal to the implanted portion were recorded serially 2–6 and 1–22 months after surgery in the rat and cat, respectively. Compound nerve action potentials (CNAPs) were also recorded in the proximal nerve following electrical stimulation of the distal nerve. In order to assess the lesion-induced alterations in the CNS pathways after operation, motor evoked potentials (MEPs) and soma-tosensory evoked potentials (SEPs) were also examined. In some animals, intra-cellular recordings were made from the lumbosacral motoneurons innervating the hind limb muscles. Details of the procedure are described elsewhere [3,4,7,8].

Histology

At 2 and 4 months (rat) and 6 and 22 months (cat) postoperatively, animals were sacrificed and transcardially perfused with 0.1 M phosphate buffered saline (PBS) prewashed followed by 0.1 M PBS (pH 7.4) containing 1% glutalaldehyde. Tissue specimens harvested from the regenerated nerve were postfixed in the 2% osmium tetroxide for 12 h and embedded in an Epon 812 resin. Serial sec-tions were cut on a microtome (1 µm) and stained with hematoxylin-eosin, tolui-dine blue for light microscopic observation. The specimens were cut at 70–90 nm using a ultramicrotome (MT-500, Sorrall) and stained with lead citrate and uranyl acetate using the Reynolds methods. Electron microscopic observation was performed using a transmission type microscope (H-7000, Hitachi).

Results and Discussion

Gelatin tube (rat sciatic nerve: 10 mm)

At 2 months postoperatively, many regenerated fibers with numerous Schwann cells were observed. At 4 months after operation, both the number and diameter of myelinated fibers were markedly greater than those at 2 months, although my-elin sheaths were still thinner compared to the normal. Electrophysiological stud-ies showed that both conduction velocities (CMAPs and CNAPs) and evoked potentials (MEPs and SEPs) attained plateau level at 4 months. Considerable return of motor performance was observed at 6 months after operation [3]. As to the rat sciatic nerve, similar results have been reported using various types of nerve guide channels and methods of evaluation [9,10]. The rodents themselves have a supreme capability of healing and regeneration compared to other higher mammals. The reverse is the case of adult human primates, especially of aged ones [9]. In addition, we could assess the gap of 10–15 mm in sciatic nerves, which is not large enough for actual clinical application. It is needless to say that the rat sciatic nerve model is useful and still plays an important role in examining new materials or methods for the regeneration of nervous tissue. However, considering that our ultimate goal in the regenerative study is to satisfy the present urgent demand for clinical application, more advanced systems are needed, which could clearly evaluate the acceleration of nerve repair across a long-distance gap.

Collagen tube (cat facial nerve: 5 mm)

We evaluated facial nerve regeneration over a 5-mm gap in adult cats, using collagen obtained from porcine skin as a biodegradable material. Collagen was chosen as the matrix of the tube given the excellent regeneration after median nerve repair in monkeys and sciatic nerve in rats. Electrophysiological study confirmed the recovery of electrical activity of the collagen-implanted regenerated nerve. Light microscopic examination of collagen-tube-implanted specimens revealed a well vascularized regenerated nerve, which under an electron microscope showed many myelinated axons surrounded by Schwann cells and unmyelinated axons. Horseradish peroxidase (HRP) staining demonstrated labeling of facial motoneurons in the brain stem and facial nerve terminals in the neuromuscular junction, also confirming of the whole facial nerve tract from the reinnervated muscles, passing through the regenerated site to the brain stem [4].

PGA-collagen tube (cat sciatic nerve: 25 mm)

In the cat sciatic nerve, we examined the nerve regeneration that occurred over a 25-mm gap using a novel biodegradable polyglycolic acid (PGA)-collagen tube which was filled with neurotrophic factors. Histological examinations carried out 4–16 months after implantation of the tube revealed regeneration of well vascularized nerve tissue. Following injection of HRP into a site peripheral to the regenerated segment of the sciatic nerves, motoneurons in the ventral horn of the spinal cord, afferent terminals in the medial portion of the dorsal column of the medulla oblongata, and sensory afferent nerve terminals in the dorsal horn of the spinal cord were labeled [7]. At 6 or 8 months after operation, toluidine blue stained section revealed many myelinated fibers with myelin sheaths and unmyelinated fibers. Electron microscopic observation showed the predominance of thick myelinated fibers in the regenerated nerves 17 months after operation, although the myelin thickness remained smaller than the one in the normal control [8]. Thin myelinated axons encapsulated with Schwann cells and scattered unmyelinated axons were also observed at the same period. Morphometric analysis showed that the mean diameter and number of the regenerated axons per area 6 months after surgery was about 70 and 130% of that of normal axons, respectively [7,8].

CMAPs and CNAPs were obtained within 1 month after operation, while MEPs and SEPs were evoked after 1–2 months. All the evoked potentials attained the plateau level in less than 4 months after operation. Functional recovery of walking movement was later than that of SEPs or MEPs. Initial flexion of the lower extremities was observed 3—4 months after operation in all operated animals. At 6 months after operation, functional recovery was attained at a considerable level, although movement in the operated hind limb was slower than the normal side and disuse atrophy of reinnervated muscles was manifested. At 8—21 months after operation, antidromic action potentials of a single motoneuron in the ventral horn of the spinal cord were evoked following stimulation of respective nerves innervating flexor or extensor muscle. This finding suggests that some of the regenerated nerve branch within the regenerated segments. On the regenerated side, stimulation of two separate muscle nerves evoked antidromic action potentials in 12 out of 26 motoneurons, while antidromic action potentials were not obtained from multiple stimulation sites in normal preparations. Although persistent innervation to both flexors and extensors by a single motoneuron may lead to impaired coordination of reciprocal muscles in walking, it may be favorable for both motoneurons and muscle fibers to maintain transport of trophic substances between them [7,11]. The most interesting observation was that animals exhibited remarkable functional returns in the chronic stage after operation despite the incomplete recoveries in electrophysiological and histological evaluations. This evidence suggests that remodeling in the CNS might significantly serve to a functional reconstruction of the PNS after peripheral nerve injury [11].

Conclusion

PGA-collagen tube is a promising tool for enhancing PNS regeneration across a long-distance gap. We must be concerned not only in developing a nerve guidance channel, but also establishing proper animal systems with the view of clinical application. Furthermore, we never hesitate to take advantage of the recent tissue engineering approaches such as the use of functionalized gel, seeding with Schwann cells and use of genetically engineered cells. Thus, we believe that these up-to-date approaches will make the best use with our biomaterial-based approaches.

Acknowledgements

This study was supported by the Grant-in-Aid for "Research for the Future Program" from Japan Society for the Promotion of Science (JSPS-RFTF96100203).

References

1. Valentini RF, Aebischer P. Strategies for the tissue engineering of peripheral nervous tissue regeneration. In: Lanza RP, Langer R, Chick WL (eds) Principles of Tissue Engineering. San Diego: R.G. Landes Company and Academic Press Inc., 1997;671—684.
2. Silver FH, Garg AK. Collagen: characterization, processing and medical applications. In:

Domb AJ, Kost J, Wiseman DM (eds) Handbook of Biodegradable Polymers, Drug Targeting and Delivery, vol 7. Amsterdam: Harwood Academic Publisher, 1997;319—346.

3. Lee G, Nakamura T, Simizu Y, Tomihata K, Ikada Y, Endo K. Experimental study of a nerve guide-tube made from dehydrothermally treated gelatin: application to repair of gap in rat sciatic nerve. Jpn J Art Organ 1993;22:364—369.

4. Kitahara AK, Suzuki Y, Peng Q, Nishimura Y, Suzuki K, Kiyotani T, Takimoto Y, Nakamura T, Shimizu Y, Endo K. Facial nerve repair using collagen nerve conduit. Scand J Plast Reconstr Surg Hand Surg 1998;(In press).

5. Kiyotani T, Nakamura T, Shimizu Y, Endo K. Experimental study of nerve regeneration in a biodegradable tube made from collagen and polyglycolic acid. ASAIO J 1995;41:657—661.

6. Kiyotani T, Teramachi M, Takimoto Y, Nakamura T, Shimizu Y, Endo K. Peripheral nerve regeneration in a PGA-collagen composite tube. Jpn J Art Organ 1996;25(2):476—480.

7. Kiyotani T, Teramachi M, Takimoto Y, Nakamura T, Shimizu Y, Endo K. Nerve regeneration across a 25-mm gap bridged by a polyglycolic acid-collagen tube: a histological and electrophysiological evaluation of regenerated nerves. Brain Res 1996;740:66—74.

8. Suzuki K, Kiyotani T, Kitahara AK, Suzuki Y, Nishimura Y, Yamamoto Y, Takimoto Y, Nakamura T, Shimizu Y, Endo K. Development of PGA-collagen channel for peripheral nerve regeneration — functional evaluation. Jpn J Art Organ 1998;27(2):490—949.

9. Mackinnon SE, Dellon L. Clinical nerve reconstruction with a bioabsorbable polyglycolic acid tube. Plast Reconstr Surg 1990;85:419—424.

10. Tong X, Hirai K, Shimada H, Mizutani Y, Izumi T, Toda N, Yu P. Sciatic nerve regeneration navigated by laminin-fibronectin double-coated biodegradable collagen grafts in rats. Brain Res 1994;663:155—162.

11. Kuno M. The Synapse: Function, Plasticity, and Neurotropism. Oxford: Oxford University Press, 1995;1—249.

JSPS TISSUE ENGINEERING PROJECT: TISSUE ENGINEERING FOR ORGAN REGENERATION

Retinal transplantation of clonal adult rat hippocampus-derived neural stem cells

Masayo Takahashi[1], Akihiro Nishida[1], Ichiro Nakano[2], Jun Takahashi[2], Akira Mizoguchi[3], Chizuka Ide[3] and Yoshihito Honda[1]
Departments of [1]Ophthalmology and Visual Sciences, [2]Neurosurgery, and [3]Anatomy, Graduate School of Medicine, Kyoto University, Kyoto, Japan

Abstract. Recently, attention has been directed toward identification and characterization of multipotent progenitor cells in the brain. Gage's group has reported that multipotent progenitor cells can be isolated from the adult hippocampus of rats. To investigate the plasticity of neuronal phenotypes that can be generated by clonal AHSC in vivo, AHSC were grafted heterotopically into the rat eye. Two clones chosen for this work carried retroviral marker genes encoding either cytoplasmic (LNPoZ) or nuclear localized β-gal (LgZnSN). 300,000 of these cells were injected into the vitreous space of the adult rat eye, or 150,000 were injected into the subretinal space or vitreous space of the newborn rat eye. Rats were sacrificed and the eyes were processed for histology 2, 4 or 8 weeks later. Four weeks after grafting into pups' eyes, the AHSC were well integrated into the retina and showed the appropriate morphology and position of amacrine, bipolar, horizontal, photoreceptor and Müller cells. This indicates that neural stem cells isolated from the adult hippocampus retain an embryonic-like plasticity in their ability to generate a wide variety of heterotypic neuronal phenotypes when presented with appropriate local cues.

Keywords: differentiation, neural stem cells, retina, transplant.

Introduction

The developing brain contains a spectrum of differentiation, from a self-sustaining population of stem cells that have the ability to give rise to all the cells of the central nervous system up until mature differentiated cells with no ability to divide. In between these two populations are progenitor cells, which are functionally immature and retain a limited proliferative capacity.

The presence of stem cells within the CNS has been a subject of great interest. Recent studies have demonstrated that neural stem cells persist in regions of the adult brain where neurogenesis continues throughout life. In 1992, Dr Weiss' group isolated EGF-responsive neural progenitor cells from the subventricular zone of an adult mouse [1]. In 1995, Dr Gage's group isolated bFGF-responsive

Address for correspondence: Masayo Takahashi MD, PhD, Department of Ophthalmology and Visual Sciences, Graduate School of Medicine, Kyoto University, Shogoin Sakyo-ku, Kyoto 606, Japan. Tel.: +81-75-751-3255. Fax: +81-75-751-3258. E-mail: masataka@kuhp.kyoto-u.ac.jp

neural progenitor cells from adult rat hippocampus [2]. Detailed analyses of clones derived from these cultures revealed that progenitors from the adult rat hippocampus have the characteristics of neural stem cells, such as the ability to self-renew and to generate both glia and neurons in vitro [3].

Isolation of adult rat hippocampal progenitors [3]

Fresh isolates of hippocampal progenitors were generated from dissociated adult rat hippocampus and cultured in DMEM/F12 supplemented with N2 and basic fibroblast growth factor. The primary culture was passaged four times and then a fraction of the resulting cells were treated with replication-defective retroviral vectors to genetically mark individual cells. Clones of retrovirally marked cells were isolated by plating the transduced cultures at limiting dilution under selection of the retrovirally transferred neomycin phosphotransferase gene and genes for cytoplasmic β-galactosidase gene. Used vectors carry a neogene and genes for either human placental alkaline phosphatase, cytoplasmic β-galactosidase, or a nuclear localized β-galactosidase. Clones carrying each of these enzyme markers were designated as AP, PZ, or Zn clones, respectively.

Characterization of isolated stem cells

In serum-free medium with bFGF, the neural stem cells are phase-bright, rounded with a few thin processes, and continue to proliferate (Fig. 1A). The majority of the proliferative cells were immunoreactive for various immature cell markers, such as nestin, and A2B5 [3]. No immunoreactivity was detected for

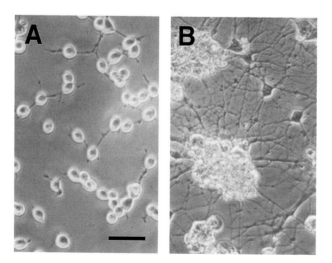

Fig. 1. Adult rat hippocampus-derived stem cells (AHSC) in vitro. **A:** When proliferating in the presence of FGF-2, the AHSC cells had round bright cell bodies with short thin processes. **B:** After 14 days in culture with retinoic acid, they extended long processes.

fibronectin; an extracellular matrix protein elaborated by fibroblast, ED-1; a marker for microglia and monocyte, and Von Willbrandt factor expressed by endothelial cells [3]. To evaluate each clone's ability to generate neurons or glia, cells were induced to differentiate by withdrawal of bFGF and addition of reti-

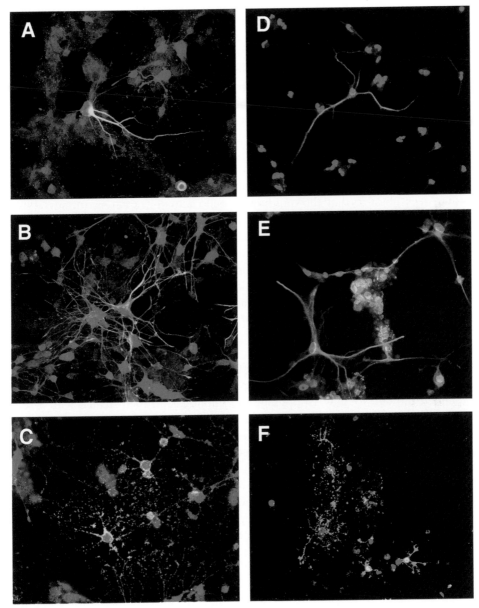

Fig. 2. Immunofluorescent staining of clonal stem cells after 2 weeks of treatment with retinoic acid. Immunoreactive cells for Map2ab (**A, D,** green); a neuronal marker, GFAP (**B, E,** green); an astrocyte marker, or GalC (**C, F,** green); an oligodendrocyte marker. These results indicated that these clones were able to generate all three lineages that exist in the central nervous system. **A, B, C:** cytoplasmic β b-gal labeled clone; PZ5. **D, E, F:** nuclear β-gal labeled clone; Zn5.

noic acid. Once bFGF is withdrawn and retinoic acid is added to the medium, the cells stop proliferating and begin to extend processes (Fig. 1B). Figure 2 shows that the clonal adult hippocampus-derived stem cells can give rise to any of all three lineages that exist in the central nervous system.

Each clone showed all the three lineages after differentiation, although the proportions of each lineage are different from clone to clone [3]. To confirm that all three lineages were derived from single cells, genomic DNA was prepared from each clone and evaluated for the presence of a single proviral band [3]. Each clone contains a single detectable proviral band and those populations that are indeed clonal display the same proviral band in DNA from proliferating or differentiated populations. This indicates that the differentiated neurons, astrocytes, and oligodendrocytes were all derived from the same marked progenitor.

Karyotyping

Although the clones were initiated directly from normal adult rat hippocampal tissues, it is possible that the multipotent cells actually represent immortalized cell lines. Rodent cell cultures frequently generate cells that escape from growth regulation as a result of genetic abnormalities accumulated in long-term culture. Overt indications of accumulated abnormalities are loss of growth control, changes in morphology and alterations in ploidy or karyotype. These clonal cells in this work showed consistent morphology and growth characteristics throughout the studies. Also, karyotype analysis revealed that the cells we used in this study were normal diploid without translocations [3].

These results demonstrate that clonal populations arising from single progenitors can give rise to multiple lineages. By definition, self-renewing stem cells should retain a multipotent phenotype through multiple rounds of replication. To demonstrate this in the hippocampal cultures, three marked clones were expanded through roughly 10 population doublings and then treated with a second retroviral vector carrying a hygromycin phosphotransferase gene. Subclones were isolated by selection in hygromycin-B. Each subclone generated was able to produce neurons, astrocytes, and oligodendrocytes, which means the mother cell could produce the same multipotent progenitor cells by replication [3].

These progenitors from the adult rat hippocampus have the characteristics of neural stem cells, the ability to self-renew and to generate both glia and neurons in vitro. Therefore, we can say that they were self-renewing stem cells and the neural stem cells can be isolated from the CNS.

Transplant of neural stem cells

When the adult hippocampus-derived stem cells were grafted into adult rat hippocampus, the cells migrated and differentiated into neurons or glia depending on their terminal site of migration [2]. Thus, it is likely that the local cues or signals influence fate choice and terminal differentiation. This study raises the question of whether the hippocampus-derived cells can undergo neuronal differentiation in other neurogenic areas of the brain. For example, if the adult hippocampal cells are grafted into the olfactory bulb, a region where neurogenesis occurs in adulthood, they can respond to persistent cues in adult brain for survival, migration and neuronal differentiation. Furthermore, the grafted cells expressed tyrosine hydroxylase in the olfactory bulb that does not exist in hippocampus [4]. This result suggests that regional cues in the adult central nervous system can direct differentiating neurons down specific phenotypic pathways.

Retinal transplant of hippocampus-derived neural stem cells [5]

The retina is a part of the central nervous system and it is often used as a good experimental model of CNS because of its relatively simple structure [6]. Light passes through the transparent retina and is captured by photoreceptors in the outer nuclear layer. The signals pass across the bipolar cells and ganglion cells. Horizontal cells and amacrine cells modify the signals. Horizontal cells lie along the outer margin of the inner nuclear layer; the bipolar cell perikarya are predominantly located in the middle of the layer, and amacrine cells are arranged along the inner border of the inner nuclear layer. The perikarya of the ganglion cells make up the most proximal layer; the ganglion cell layer. The predominant type of glial cell in the retina is called the Müller cell. Müller cells extend vertically through the retina from the distal margin of the outer nuclear layer to the inner margin of the retina. Other types of glial cells, such as astrocytes, are observed in the inner part of the retina (Fig. 3 right). It is well known that all these cells including photoreceptor and Müller cells were differentiated from the common progenitor cells [7].

Methods

Cultured stem cells were harvested with trypsin, washed with PBS, and suspended at a density of 10^5 cells per µl in Dulbecco's PBS and 20 ng of bFGF per µl. A 10-µl Hamilton syringe with a 30-gauge beveled needle was used to slowly inject 3 µl of cells into the vitreous cavity of anesthetized 2-month-old adult rats or 1.5 µl into neonatal rats. After 2, 4, and 8 weeks, eyes were fixed in 4% paraformaldehyde and cut at 20-mm-thick sections in a cryostat.

The grafted stem cells express cytoplasmic β-galactosidase, so they can be identified using anti-β-galactosidase antibody (Fig 4A, B). Two weeks after grafting into pups' vitreous cavity, most cells attached along the inner surface of the retina (Fig. 4A), and 4 weeks after injection, many surviving cells were well integrated within the retina (Fig. 4B). The results at 8 weeks were similar to those at the 4-week point, indicating that transgene expression remained strong for at least 8 weeks, the latest time point evaluated. Figure 4D shows the proportions of β-Gal-immunoreactive cells in various layers of the retina. The integration patterns of the two clones, the cytoplasmic-labeled clone PZ5 and the nuclear-

82

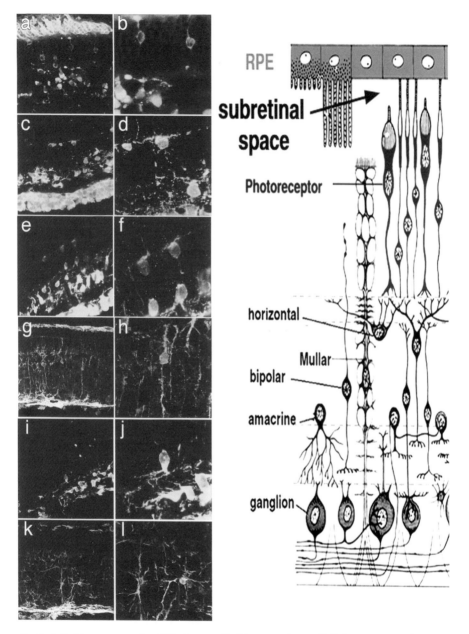

Fig. 3. β-Gal-immunoreactive PZ5 clone grafted into pups' eyes. Cells in a, c, e, g, i, k are shown at higher magnification in b, d, f, h, j, l. b) photoreceptor; d) horizontal, f) bipolar; h) Müller; j) amacrine; l) astrocyte-like cells.

Sections were stained with mouse anti β-Gal diluted 1:5,000 (Jackson Immunochemicals) and double incubated (g, h, k, l) with guinea pig antiglial fibrillary acidic protein (GFAP) diluted 1:500 (Advanced Immunochemical). The primary antibody was detected with fluorescein isothiocyanate (FITC)- and Texas red-labeled secondary antibodies (Jackson Immunochemicals, West Grove, PA; 1:250 in PBS/TS).

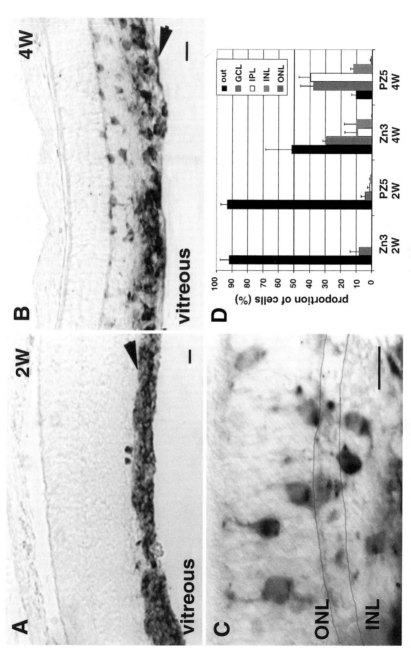

Fig. 4. β-Gal immunoreactive AHSCs grafted in pups' eyes at 2 (**A**) and 4 (**B**) weeks after implantation. Arrowheads indicate the inner surface of retina. Bar = 10 mm. **C**: β-Gal-immunoreactive cells that have proper morphology and position for photoreceptor cells. **D**: Proportions of β-Gal-immunoreactive cells in various layers of the retina. out; cells attached at inner surface of the retina. GCL; retinal ganglion cell layer, IPL; inner plexiform layer, INL; inner nuclear layer, ONL; outer nuclear layer.

labeled clone Zn3 were similar. At the 2-week time point most of the cells existed at the inner surface of the retina, and at the 4-week time point grafted cells migrated into as far as the inner nuclear layer. A small amount of cells migrated into the outer nuclear layer.

Control experiments

Control grafts of β-Gal-marked skin fibroblasts did not integrate into the retina, but rather attached to the retinal surface and caused retinal detachment. Although these cells were in intimate contact with the retinal surface, no β-Gal immunoreactivity was transferred to the host retinal cells. In addition, injection of 1,000 times more purified *Escherichia coli* β-Gal enzyme than was contained in grafted stem cells into the pups' vitreous cavity or into adult cerebral cortex showed no β-Gal uptake by host neurons or glia 3 h, 1 day or 3 days after injection. Similarly, injection of freeze-thawed stem cells into the vitreous space demonstrated that enzyme leak from damaged or from dead cells was not taken up by host retinal cells. These controls strongly support the contention that all β-Gal-immunoreactive cells within the retina were derived from the grafted cells.

Notably when the stem cells were grafted into the vitreous cavity of 2-month-old adult rats, no cells were observed which integrated into the retina at the 4-week point.

Differentiation of grafted stem cells

The cells that integrated into the inner nuclear layers of the retina showed a strikingly similar morphology and position to Müller, amacrine, bipolar, and horizontal cells (Fig. 3a-l, 4C). Retinal slices were stained for β-Gal and a variety of retinal cell markers including calbindin, tyrosine hydroxylase, Thy-1.1, Ret-P1, S100β, and GFAP. Some grafted cells that showed Müller-like morphology expressed S100β but not GFAP (Fig. 3h). This is compatible with retinal glia but not with astrocyte in the brain, so that it appears that the grafted cells differentiated into the retinal-specific glia. As for the neuronal markers, grafted cells showed no retinal-specific markers even though they showed general neuronal markers such as Map2 or Map5, however, they possessed perfect morphology and localization as retinal neurons. Whether the lack of retinal neuronal markers reflects incomplete differentiation due to a lack of cues present in early retinal development or an intrinsic limitation of hippocampus-derived cells remains to be determined.

Transplant into adult retina

When the stem cells were grafted into the pups' eyes, they migrate and integrate well in the retina. However, when the stem cells were injected into adult eyes, they never migrated into the retina even 4 weeks after injection. If the stem cells could not integrate into adult retina, it is impossible to use them for clinical treatment. Now we have the preliminary data on the transplantation of neural stem cells into adult rat retina [8]. We observed that they could migrate into injured adult retina and differentiated into glia and neurons. They were distributed around the injury site, where Müller cells were activated. Their migration may be in response to the local cues that were secreted from the activated Müller cells.

Summary and prospects

The adult rat hippocampus-derived neural stem cells reacted to the heterotopic environmental cues to differentiate into neuron or glia. They possessed appropriate morphology and position to the retinal-specific neurons and glias. However, some factors are missing that make them fully differentiated into retinal neurons. It may be an extrinsic factor that lacks in the retina after birth or it may be an intrinsic factor. Nevertheless, surprisingly good migration and integration of the stem cells into the retina, compared to other studies [9—11], show the possibility of usage of stem cells for retinal neural transplant.

Presently, there is no treatment when the retinal neurons are damaged and the patients become blind. Neural transplant might be the only way to treat such a condition. Some groups have begun clinical trials of transplant of embryonic retina for a retinal degenerative disease called retinitis pigmentosa in recent years [12]. Although even if it is useful, the limit of transplantation of embryonic tissue is clear due to the imbalance between supply and demand. Research on neural stem cells is still in the beginning stages. However, it has provided the basis for the new therapeutic strategies. The use of neural stem cells makes it possible to obtain engineered cell lines with homogenous clonal properties. They integrated into the central nervous system, allowing close contact with host cells. These characteristics allow us to use them as a vehicle for ex vivo gene therapy or neural transplant into CNS. In the future, well-characterized human neural stem cells will become the key to developing clinically applicable neuroregenerative treatment.

References

1. Weiss S, Reynolds BA, Vescovi AL, Morshead C, Craig CG, van der Kooy D. Is there a neural stem cell in the mammalian forebrain? Trends Neurosci 1996;19:387—393.
2. Gage FH, Coates PW, Palmer TD, Kuhn HG, Fisher LJ, Suhonen JO, Peterson DA, Suhr ST, Ray J. Survival and differentiation of adult neuronal progenitor cells transplanted to the adult brain. Proc Natl Acad Sci USA 1995;92:11879—11883.
3. Palmer TD, Takahashi J, Gage FH. The Adult rat hippocampus contains primordial neural stem cells. Molec Cell Neuroscience 1997;8:389—404.
4. Suhonen JO, Peterson DA, Ray J, Gage FH. Differentiation of adult hippocampus-derived progenitors into olfactory neurons in vivo. Nature 1996;383:624-627
5. Takahashi M, Palmer TD, Takahashi J, Gage FH. Widespread integration and survival of adult-derived neural progenitor cells in the developing optic retina. Molec Cell Neuroscience 1997;(In press).

6. Dowling JE. Retina. An Approachable Part of the Brain. Harvard: Belknap, 1987;12—33.
7. Turner DL, Cepko CL. A common progenitor for neurons and glia persists in rat retina late in development. Nature 1987;328:131.
8. Nishida A, Takahashi M, Nakano I, Takahashi J, Mizoguchi A, Ide C, Honda Y. Clonal adult hippocampus-derived stem cells can migrate into injured adult rat retina. Invest Ophthalmol Vis Sci 1998;39:s19.
9. Seiler MJ, Aramant RB. Transplantation of embryonic retinal donor cells labeled with BrdU or carrying a genetic marker to adult retina. Exp Brain Res 1995;105:59—66.
10. Gouras P, Du J, Kjeldbye H, Kwun R, Lopez R, Zack DJ. Transplanted photoreceptors identified in dystrophic mouse retina by a transgenic reporter gene. Invest Ophthalmol Vis Sci 1991; 32:3167—3174.
11. Trisler D, Rutin J, Pessac B. Retinal engineering: Engrafted neural cell lines locate in appropriate layers. Proc Natl Acad Sci USA 1996;93:6269—6274.
12. Kaplan HJ, Tezel TH, Berger AS, Wolf ML, Del Priore LV. Human photoreceptor transplantation in retinitis pigmentosa. A safety study. Arch Ophthalmol 1997;115:1168—1172.

Scaffold structure for a bioartificial liver support system

J. Mayer[1], E. Karamuk[1,2], K. Interewicz[1], T. Akaike[2] and E. Wintermantel[1]

[1]*Biocompatible Material Science and Engineering, Department of Materials, Swiss Federal Institute of Technology, Zurich, Switzerland; and* [2]*Faculty of Bioscience and Biotechnology, Tokyo Institute of Technology, Tokyo, Japan*

Abstract. In this study we introduced a composite scaffold proposed for bioartificial liver support systems based on the principles of textile superstructuring. A woven PET fabric was coated with various biodegradable PLGA/NaP film systems on one side in order to obtain a polarized structure and combined with biomimetic polymers to provide receptor-mediated cell attachment sites. While the 85:15 PLGA showed no significant weight loss during the observed immersion time, the 50:50 PLGA matrix exhibited a weight loss of 50%, starting on the 8th day and increasing in rate. The GPC data for scaffolds of 50:50 PLGA shows, that the molecular weight decreased to less then 8,000 g/mol on the 8th day. The composite structure ensured the stability of the membrane during degradation. Optimal conditions for culturing of primary mouse hepatocytes were found using large, PVLA (100 µg/ml) coated meshes combined with EGF (50 ng/ml) leading to aggregates of up to 160 µm in size after 48 h incubation. A new method was developed to measure simultaneously the diffusion behavior of different molecules from 0.2 to 70 kDa in these scaffold systems. Perfusion properties were investigated from the physiological hydrostatic pressure range up to laminar/turbulent transition.

Keywords: biofunctionalization, composites, degradables, diffusion, perfusion, scaffolds.

Introduction

Biocompatibility of materials has been defined as compatibility of a technical with a biological system [1]. Biocompatibility is described with structural compatibility and with surface compatibility. Textile superstructures for tissue engineering scaffolds have been proposed [2,3] for their ability to provide optimal spatial and nutritional conditions for cell maintenance by the arrangement of structural elements, such as pores and fibers. They are considered to support extracellular matrix (ECM) formation, mimicking the target tissue.

In contrast to potential kidney recipients, patients in liver failure do not have similar alternatives, such as dialysis, until an organ becomes available for transplantation. Therefore, there is a critical need for techniques which provide long-term, as well as short-term liver support. An effective artificial liver support system should be capable of carrying out the liver's essential processes, such as synthetic and metabolic functions, detoxification, and excretion [4,5]. Dialyzer-like

Address for correspondence: Dr J. Mayer, Biocompatible Materials Science and Engineering, Wagistrasse 23, CH 8952 Schlieren, Switzerland. Tel.: +41-16336311. Fax: +41-16331120. E-mail: mayer@biocomp.mat.ethz.ch artificial liver systems which replace single liver functions lack the metabolic and synthetic functions of the liver [6—8] and the survival rate of patients with fulminate hepatitis does not exceed 30% [4,8]. Therefore, the use of isolated hepatocytes as the basis for artificial liver support, e.g., implantable hepatocyte systems and extracorporal hepatocyte bioreactors, so-called bioartificial liver support systems (BAL, Table 1), is envisaged to replace a broader spectrum of liver-specific metabolic functions. In both approaches, isolated liver cells are used in a variety of configurations, e.g., suspensions [5,9], attached to substrates [10—12] and encapsulated in microspheres [4,13] or hollow fiber systems [14—16].

BAL devices support liver functions through an external circuit that exchanges plasma components with the patient. In comparison to implantable systems, extracorporal bioreactors have some advantages [17,18]. Bioreactors allow better control of the cellular environment, particularly with respect to transport problems. Oxygen delivery and macromolecular exchange can be engineered. In contrast, implanted systems must rely on the characteristics of the surrounding milieu. In addition, immunological rejection is less problematic because of the potential to separate the recipient's lymphocytes away from the exogenous hepatocytes by plasmapheresis. Disadvantages of an extracorporal system include the need for vascular access and the associated potential of thromboembolic complications.

Setup of the scaffold systems

In this study, a composite scaffold is proposed for liver cell engineering. The scaffold consists of a woven polymer fabric with well-defined pores which is coated on one side with a biodegradable polymer matrix. Figure 1 illustrates the working principle aiming to mimic a functional unit of the liver and proposed membrane composite system. Thin, porous films have promising characteristics for tissue engineering because their degradation rate, porosity, and surface chemistry can be controlled by the fabrication process. Combining them with fibrous structures

Table 1. Examples of extracorporal bioartificial liver systems based on hollow-fiber bioreactors.

Author	Bioreactor design	Observed effects
Wolf and Munkelt 1975 [14]	Cells are cultivated on the exterior surfaces of semi-permeable capillary hollow fiber membranes to be perfused with the host animals blood	Bilirubin uptake and conjugation
Moscioni et al. 1990 [12]	Cryopreserved human hepatocytes. Microcarrier-attached cells in the hollow-fiber extracapillary space	Cyclosporine metabolism
Shiraha et al. 1995 [19]	Multicellular spheroids encapsulated in microdroplets of 5% agarose in the extracapillary space of a hollow-fiber module	Albumin synthesis and amino acid removal
Wu et al. 1995 [15]	Porcine hepatocytes spheroids entrapped in a three-dimensional collagen matrix in the interluminar space of a hollow-fiber column	Albumin synthesis, ureagenesis and P450 cytochrome activity

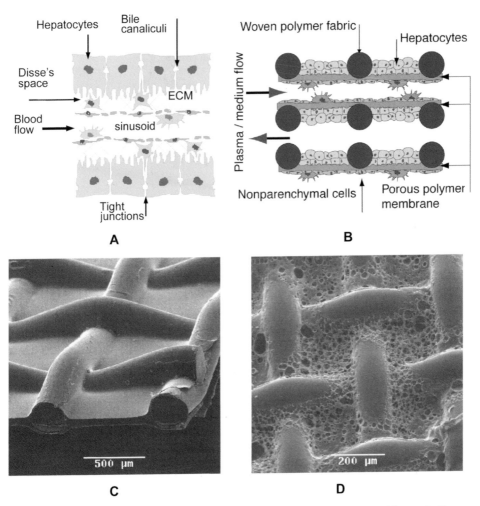

Fig. 1. Working principle and setup of the composite scaffold system. **A:** Functional liver unit; **B:** proposed polarized scaffold setup aiming at co-culturing of hepatocytes and nonparenchymal cells. **C:** Cross-section view of a 720-μm mesh scaffold showing the integration of the 20- to 50-μm-thick polymer film being attached on one side of the fabric. **D:** 200-μm mesh scaffold with highly porous membrane by leached NaP particles (6 h at 50°C).

such as fabrics opens the possibility for applications in larger scales, e.g., in vivo and in bioreactors. In Table 2 some specific properties of the woven fabric and the polymer film are shown that are combined in a composite scaffold.

We assume that the three-dimensional structure of the scaffold will influence the functionality of hepatocytes and the generation of a tissue-like architecture in vitro, as it was observed for uncoated nylon fabrics [20] and in collagen-sandwich configurations [17]. The degradable polymer matrix can be combined with artificial glycopolymers such as PVLA. They have been shown to influence

Table 2. Properties of the woven fabric and the polymer film that are combined in a composite scaffold.

Woven polymer fabric	Polymer film matrix
Stiffness	Microscopic porosity
Macroscopic porosity	Permeability
Macroscopic surface topology	Cellular affinity by receptor-specific ECM coating
Structural unit	Delivery system
Oxygen and nutrients supply through woven hollow fibers	

hepatocyte attachment and function, since they carry concentrated groups of carbohydrates which are recognized by specific protein receptors at the cell membrane [21,22]. The degradable polymer film will act as a membrane, separating two different environments and thus creating a gradient of the concentration of components in the medium and of partial pressure. We hypothesize that this form of mechanical stimulus will positively influence the faith of cell cultures on either side of the film and will contribute to the development of liver-like tissues.

A two-phase casting technique was applied to coat woven PET fabrics of 50- to 800-μm mesh width on one side with a thin (< 100 μm) PLA/PGA film (Fig. 1). Therefore, we investigated the following matrix systems (Table 3).

Properties of scaffold composites

The composite scaffolds exhibit a polarized structure, with a plain "film side" and a highly organized "fabric side" with well-defined pores between 800 and 50 μm depending on the mesh size of the fabric. The film is anchored mechanically between the fibers (Fig. 1).

To study in vitro degradation of the scaffold matrix, square pieces of 15×15 mm were allowed to degrade in 12-well culture plates in phosphate buffered saline (PBS (–)) at 37°C. The medium was replaced every 4 days. The samples were washed in deionized water and lyophilized for 24 h before being subjected to various analyses to detect weight loss, water absorption, and molecular weight decrease.

Table 3. Investigated matrix systems.

	Molecular weight (kDa)	PLDLA	PGA	Typ	Filler
PLA/PGA copolymer	200	85%	15%	RG858	—
PLA/PGA copolymer	20	50%	50%	RG503	—
Na$_2$HPO$_4$ filled	200	85%	15%	RG858	10%
Na$_2$HPO$_4$ filled	200	85%	15%	RG858	33%
Na$_2$HPO$_4$ filled	200	85%	15%	RG858	50%

91

Out of tested polymers, only the 50:50 PLGA showed relevant degradation in the observed period (Fig. 2). The films in these scaffolds lost 90% of their initial molecular weight after 14 days and 50% of their initial mass after 25 days (data not shown). Initial matrix degradation was observed in regions with high surface axial strain at the film/fabric interface. However, even if the matrix had lost its mechanical properties the structural polarity was maintained and no detachment of the film occurred. Due to the small thickness of the polymer film we did not observe a biphasic, bulk degradation behavior. Surface roughness increased considerably.

The incorporation of sodium phosphate particles (NaP) in the slow degrading 85:15 PLGA matrix led to a significant increase in water uptake and porosity but did not change the degradation rate (Fig. 3). The molecular weight was, on the contrary, observed to decrease slower in this case. The reduced degradation rate may be a consequence of the alkaline properties of the Na_2HPO_4 salts, which would be expected to disrupt autocatalytic degradation. The increase in porosity due to salt leaching resulted in an inhomogeneous pore distribution. No significant influence of incorporated salt particles on the pH decrease was observed (data not shown). We did not observe an influence of the fabric size over the degradation behavior (Fig. 2). Generally, scaffolds with larger fabrics had a more uniform film thickness.

A blend of low molecular weight (20 kDa) PLGA (50:50) and high molecular

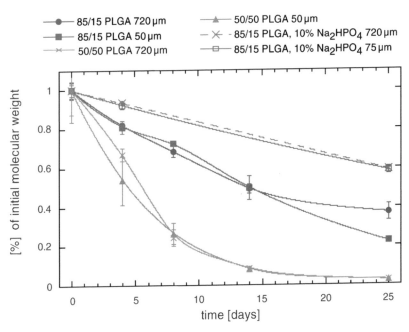

Fig. 2. Degradation behavior of PLGA scaffold systems illustrating the influence of mesh size and NaP content.

92

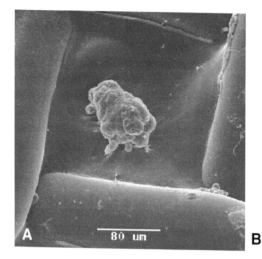

Culture conditions	Mesh size (µm)		
	720	200	50
EGF(+)	72	96	96[a]
PVLA (1 µg/ml)	>96	>96	–
PVLA (1 µg/ml)/EGF(+)	48	96	96[a]
PVLA (100 µg/ml)	>96	>96	–
PVLA (100 µg/ml)/EGF(+)	48	96	96[a]

> : cells show a clear tendency to aggregate formation, but not yet finished. [a]Aggregate formation, but cells spread over several mesh cavities.

Fig. 3. Aggregation of hepatocytes. In the presence of 50 ng/ml EGF, hepatocytes form spheroidal aggregates after 48 h (720 µm). No effect from the PVLA coating. Aggregate diameter did not exceed ~ 160 µm. **A:** SEM image showing an individual agglomerate on a 200-µm mesh scaffold after 96 h (EGF+, 1 µg/ml PVLA). **B:** Influence of scaffold configuration on aggregation behavior.

weight (200 kDa) PLGA (85:15) was cast on the 720- and 75-µm weave. With the use of a polymer blend matrix a uniform porosity could be achieved which resulted in a high water uptake of up to 50%. The blend matrix exhibited a weight loss as a result of the degradation of the constituent polymers (data not shown). In some cases an open porosity was observed, which is considered to be important for the permeability and mass transfer characteristics of the membrane.

One possible reason for this high porosity is a phase separation between the two components. Since the copolymer with the high PGA content is more difficult to dissolve in chloroform, it is possible, that small clusters of PGA chain segments remain in the solution leading to inhomogeneity in the matrix composition.

In vitro behavior

Co-culturing of hepatocytes seems necessary for the development of liver-like tissues in vitro [20,21,23]. The proposed scaffold system offers the possibilities of co-culturing hepatocytes and nonparenchymal cells; hepatocytes on one and nonparenchymal cells on the other side of the membrane (Fig. 1) in a perfusion culture setup. With the use of a porous degrading membrane it may be possible to create an "artificial Disse's space", where ECM proteins that are excreted by the nonparenchymal cells slowly replace the degrading polymer membrane.

To obtain a receptor-specific attachment of hepatocytes part of the PLA/PGA scaffolds were coated with PVLA, at concentrations of 1 and 100 µg/ml. After washing, the scaffolds were treated with 0.1% BSA solution (37°C, 30 min) to reduce unspecified cell attachment. Each scaffold was added to a well of a 24-well culture plate. Freshly isolated primary mouse hepatocytes were seeded on the scaffolds at a density of 50,000 cells/cm^2 (600-µl serum-free WE medium). Cultures were incubated for 4 h for initial cell attachment. Thereafter, the medium was replaced by 500-µl fresh medium with or without 50 ng EGF/ml. The medium was replaced daily.

Hepatocytes attached to the PLGA surfaces. Only in the presence of epidermal growth factor (EGF) was cell aggregation observed. The different mesh sizes had an influence on the kinetics of aggregate formation. In the case of larger pore sizes, spheroids were formed within 48 h (Fig. 3), whereas complete aggregate formation required 4 days on the finer meshes. This may be explained by the complete isolation of a limited number of hepatocytes in the case of the large meshes. This effect could be used for fast and efficient spheroid assembly, avoiding possible damage to the cells as is the case for conventional suspension techniques [25,27].

The hepatocytes on the scaffold produced albumin once complete spheroid formation was finished and the formation of the spheroids was completed. The pores of the fabric were filled with aggregates, illustrating the possibility of immobilizing metabolically highly active aggregates on textiles. These results have to be judged as preliminary; however, they support the hypothesis that our scaffold is based upon.

Perfusion and diffusion behavior

Designing a bioreactor for the final BAL system requires exact knowledge of the mass transport properties of the scaffold in order to optimize the supply of nutrients as well as the exchange of metabolites. Perfusion and diffusion are the main transport phenomena involved in the system and they set the limitations to the scaling up of the scaffold into a bioreactor.

A method was developed to measure simultaneously the diffusion behavior of molecules from 0.1 to 70 kDa. Three molecules with different UV absorption peeks were chosen as tracer materials (Table 4), which allows a simultaneous measurement and decoupling of the tracer concentrations. For each tracer molecule, UV absorption was measured separately at different concentrations. The data was linearly fitted and the coefficients were assembled into a matrix which allowed to decouple the measured absorption spectra (Eqn. 1).

$$\begin{bmatrix} A_{278}^{Albumin} & A_{278}^{Methyleneblue} & A_{278}^{Sodium-di-chromate} \\ A_{372}^{Albumin} & A_{372}^{Methyleneblue} & A_{372}^{Sodium-di-chromate} \\ A_{665}^{Albumin} & A_{665}^{Methyleneblue} & A_{665}^{Sodium-di-chromate} \end{bmatrix} \cdot \begin{bmatrix} C_{Albumin} \\ C_{Methyleneblue} \\ C_{Sodium-di-chromate} \end{bmatrix} = \begin{bmatrix} A_{278}^{Total} \\ A_{372}^{Total} \\ A_{665}^{Total} \end{bmatrix}$$

Scaffold diffusion coefficients were obtained with the time-lag method [24] and by measuring the time-dependent concentration change in steady-state condi-

Table 4. UV/Vis tracer substances, their molecular weight and UV absorption maxima.

UV tracer	Molecular weight	Main absorption peaks (nm)	
Sodium dichromate	298.05	372	280
Methylene blue	319.84	665	295
Albumin	~ 70000	278	—

tions (Fig. 4A). In scaffolds with porous matrices fabricated by salt leaching (see Fig. 1D), the diffusion of the different tracer molecules was correlated to the initial NaP content of the matrix by an exponential law (Fig. 4B).

Perfusion was studied from the physiological range of pressure gradients from

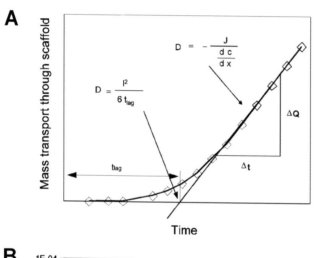

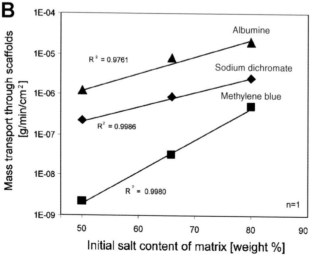

Fig. 4. Diffusion properties of the scaffold systems. **A:** Determination of the diffusion coefficient from lag time or slope. **B:** Influence of membrane porosity on mass transport.

10 mm H$_2$0 up to laminar/turbulent transition which sets a limit to the accuracy of the measurements. In the first investigations, systems of stacked fabrics of different mesh sizes were measured for their perfusion behavior. The specific perfusion through stacked fabrics decreased with an increasing number of layers, leading to the interpretation that a critical number of stacked fabrics act as a fluid-dynamical diffusor. Thus, the perfusion coefficient becomes independent from the original mesh size (Fig. 5A). In scaffolds with porous matrices the perfusion rate was correlated to the initial NaP content of the matrix by an exponential law (Fig. 5B). The measurements were performed at the physiological sinusoidal pressure of 10 mm H$_2$O [26].

Conclusion

In this study we showed the importance of matrix engineering to achieve specific cell functionality in tissue engineering, i.e., hepatocyte aggregation influenced by the three-dimensional structure of the scaffold. The two-phase casting tech-

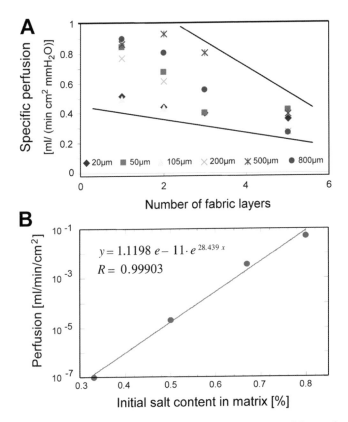

Fig. 5. Perfusion properties of fabric scaffold. **A:** Influence of the number of stacked fabrics, indicating diffusor effects. **B:** Influence of the membrane porosity on perfusion at 10 mm H$_2$O.

nique provides a mechanically stable, polarized structure, with a plain "film side", and a highly organized "fabric side". Pore sizes between 800 and 50 μm were defined by the mesh size of the fabric.

In vitro degradation was initiated at the film/fabric interface in regions with high axial surface strain. The systems maintained their structural integrity and no detachment of the film occurred. No influence of the mesh geometry on the degradation behavior was observed. The addition of water-soluble sodium phosphate particles allowed the control of the open porosity, and thereby the perfusion and diffusion properties over a wide range.

In the cell culture studies it could be shown that PVLA leads to receptor-specific attachment of hepatocytes, which we assume is necessary for stable cultures on degrading polymer surfaces. However, no aggregation was observed without the presence of EGF. The different mesh sizes had an influence on the kinetic of aggregates formation.

Further studies have to concentrate on a optimization of the physical properties of the scaffold. This includes a better control of the pore arrangement at a higher pore content as well as a higher mechanical stability and more uniform film thickness. Metabolic activity and stability of the cultured hepatocytes as well as the application of co-culturing techniques with nonparenchymal cells have to be investigated.

Acknowledgements

The authors would like to express their gratitude to everyone that supported this research: H. Ise, S.-H. Kim, R. Takei at the Faculty of Bioscience and Biotechnology, T.I.T. Tokyo and B. Eder-Aebersold, S. Ritter, K. Ruffieux, and R. Reber at the Chair of Biocompatible Material Science and Engineering, ETH Zurich. The financial support from the Tokyo Institute of Technology and the Swiss Federal Institute of Technology is greatly appreciated.

References

1. Wintermantel E, Ha S-W. Biokompatible Werkstoffe und Bauweisen. Implantate für Medizin und Umwelt. Berlin: Springer, 1996;6.
2. Wintermantel E et al. Tissue engineering scaffolds using superstructures. Biomaterials 1996;17:83—91.
3. Wintermantel E et al. Tissue engineering supported with structured biocompatible materials: goals and achievements. In: Speidel MO, Uggowitzer P (eds) Materials in Medicine. Zurich: VCH Press, 1998;3—138.
4. Kasai M et al. Is the biological artificial liver clinically applicable? A historic review of biological artificial liver support systems. Artif Organs 1994;18:348—354.
5. Dixit V. Development of a bioartificial liver using isolated hepatocytes. Artif Organs 1994;18:371—384.
6. Malchesky P. Nonbiological liver support: historic overview. Artif Organs 1994;18:342—347.
7. Yu TJ et al. Bicomponent vascular grafts consisting of synthetic absorbable fibers. I. in vitro study. J Biomed Mater Res 1993;27:1329—1339.

8. Kamlot A et al. Review: artificial liver support systems. Biotechnol Bioeng 1996;50:382–391.
9. Dixit V et al. Hepatocyte immobilisation on PHEMA microcarriers and its biologically modified forms. Cell Transplant 1992;1:391–399.
10. Mooney DJ et al. Biodegradable sponges for hepatocyte transplantation. J Biomed Mater Res 1995;29:959–965.
11. Wintermantel E et al. Angiopolarity of cell cariers: directional angiogenesis in resorbable liver cell transplantation devices. In: Steiner R, Weiss B, Langer R (eds) Angiogenesis, Key Principles: Science, Technology, Medicine. Basel, Switzerland: Birkhäuser Publishers, 1992;331–334.
12. Moscioni AD et al. Long-term cryopreserved human hepatocytes maintain drug metabolizing ability. Surg Forum 1990;41:3–4.
13. Cai A et al. Development and evaluation of a system of microencapsulation of primary rat hepatocytes. Hepatology 1989;10:855–860.
14. Wolf CFW, Munkelt BE. Bilirubin conjugation by an artificial liver composed of cultured cells and synthetic capillaries. Trans Am Soc Art Int Org 1975;21:16–26.
15. Wu FJ et al. Entrapment of hepatocyte spheroids in a hollow fiber bioreactor as a potential bioartificial liver. Tissue Eng 1995;1:29–40.
16. Mooney DJ et al. Stabilized polyglycolic acid fibre-based tubes for tissue engineering. Biomaterials 1996;17:115–124.
17. Taguchi K et al. Development of a bioartificial liver sandwich-cultured hepatocytes between two collagen gel layers. Artif Organs 1996;20:178–185.
18. Langer R, Vacanti JP. Tissue engineering. Science 1993;260:920–926.
19. Shiraha H et al. Improvement of serum amino acid profile in hepatic failure with the bioartificial liver using multicellular hepatocyte spheroids. Biotechnol Bioeng 1996;50:416–421.
20. Naughton BA et al. Sterotypic culture systems for liver and bone marrow: evidence for the development of functional tissue in vitro and following implantation in vivo. Biotechnol Bioeng 1994;43:810–825.
21. Kobayashi A et al. Regulation of differentiation and proliferation of rat hepatocytes by lactose-carrying polystyrene. Artif Organs 1992;16:564–567.
22. Akaike T et al. New frontier of biomimetic glycoengineering for cellular and tissue engineering. In: Ogata N et al. (eds) Advanced Biomaterials in Biomedical Engineering and Drug Delivery systems. Tokyo: Springer, 1996;173–178.
23. Akaike T et al. Design of hepatocyte-specific extracellular matrices for hybrid artificial liver. Gastroenterol Jpn 1993;28:45–52.
24. Barrer RM. Permeation, diffusion and solution of gases in organic polymers. Trans Faraday Soc 1939;35:628
25. Wu FJ et al. Efficient assembly of rat hepatocyte spheroids for tissue engineering applications. Biotechnol Bioeng 1996;50:404–415.
26. Arias IM. The Liver: Biology and Pathology, 3rd edn. New York: Raven Press, 1994.
27. Jaregui HO et al. The use of microcarrier roller bollte culture for large-scale production of porcine hepatocytes. Tis Eng 1997;3:17–25.

JSPS TISSUE ENGINEERING PROJECT: BIOMATERIALS FOR TISSUE ENGINEERING

Novel manipulation technology of cell sheets for tissue engineering

Masayuki Yamato[1], Akihiko Kikuchi[1], Shinichi Kohsaka[2], Tetsuya Terasaki[3], Horst A. von Recum[4], Sung Wan Kim[4,5], Yasuhisa Sakurai[1] and Teruo Okano[1]

[1]*Institute of Biomedical Engineering, Tokyo Women's Medical University, Tokyo;* [2]*Department of Neurochemistry, National Institute of Neuroscience, Tokyo;* [3]*Department of Pharmaceutics, Tohoku University, Miyagi, Japan;* [4]*Department of Bioengineering, and* [5]*Center for Controlled Chemical Delivery, University of Utah, Salt Lake City, Utah, USA*

Abstract. We have focused on novel cell manipulation techniques as a key technology in tissue engineering. We have developed temperature-responsive polymer-grafted cell culture surfaces that are hydrophobic at 37°C and change to hydrophilic below 32°C. Various cell lines adhere, spread, and proliferate on the grafted surfaces similarly to those on ungrafted commercial tissue culture dishes. By reducing culture temperature, cells are spontaneously liberated only from the grafted surfaces without the need for typical enzymatic digestion. Highly trypsin-susceptible cells such as hepatocytes and retinal pigmented epithelial cells retained higher activities of specific cell functions after recovery by reducing temperature. Confluent cells are recovered from the grafted surfaces as a single contiguous monolayer sheet with intact cell-cell and cell-extracellular matrix junctions. Therefore, viable cell sheets can be transferred from temperature-responsive surfaces to other surfaces of culture dishes or devices. With two-dimensional cell sheet manipulation, we confirmed the apical-basal localization of specific cell membrane proteins. Finally we have three-dimensional cell sheet manipulation to reconstruct tissue architectures from cell sheets, since several cell types are co-organized into defined cell sheet layers in natural tissues. Overlaying two monolayer sheets of hepatocytes and endothelial cells obtained from polymer-grafted culture surfaces provides a viable construct for in vitro fabricated liver lobule-like tissue. Cultured lamellar cell sheets preserve each cell phenotype and basic cell functions. We believe that these two- and three-dimensional cell manipulation techniques will become new revolutionary tools for tissue engineering.

Keywords: cell transplantation, co-culture, hybrid artificial organ, poly(*N*-isopropylacrylamide), temperature-responsive.

Introduction

Science and technology are advancing beyond cell and genetic engineering toward the tissue engineering age, although concepts in tissue engineering are relatively new. To attain the true goals promised by tissue engineering, further development of core materials and biological technologies is required. In the case of cell engineering, a few revolutionary techniques including cell fusion real-

Address for correspondence: Teruo Okano, Institute of Biomedical Engineering, Tokyo Women's Medical University, 8-1 Kawada-cho, Shinjuku, Tokyo 162-8666, Japan. Tel.: +81-3-3353-8111 ext. 30233. Fax: +81-3-3359-6046. E-mail: tokano@lab.twmc.ac.jp

izing the impact of monoclonal antibody mass production, automated cell sorting, as well as genetic transformation are core technologies. Other significant methods, including recombinant DNA manipulation and the polymerase chain reaction (PCR) have been developed to progress genetic engineering. Tissue engineering requires a harmonized integration of cell biology, molecular biology, and scaffolding biomaterials. One practical shortcoming is an inability to co-culture cells into tissues and to combine or manipulate separately cultured cells as integrated living structures. To achieve this technology for tissue engineering, we have focused on novel cell manipulation techniques.

We have recently developed a new cell culture dish surface that responds reversibly and dynamically to temperature changes [1]. The temperature-responsive polymer, poly(N-isopropylacrylamide) (PIPAAm) [2,3], was covalently attached to solid surfaces by specific chemical immobilization reactions [4] or electron beam irradiation [1]. Briefly, a specified amount of N-isopropylacrylamide solution in 2-propanol was spread onto tissue culture polystyrene (TCPS) surfaces. The surface was then irradiated with 0.25 MGy electron beam to graft and polymerize PIPAAm simultaneously. PIPAAm graft densities are easily controlled by changing monomer solution concentration.

This surface shows similar hydrophobicity as normal TCPS dishes above PIPAAm's lower critical solution temperature (LCST) of 32°C, where PIPAAm chains are dehydrated, collapsed, and compact on the surface. When the temperature is reduced below the LCST, grafted PIPAAm molecules rapidly hydrate and the surface becomes hydrophilic. The hydrophilic/hydrophobic surface property alteration can be tailored by choosing appropriate monomer mixtures and concentrations in surface preparation. We have utilized this intelligent material surface for cell manipulations in tissue engineering as described in the following sections.

Single cell attachment and detachment

We have succeeded in the culture of various cell types including fibroblasts, endothelial cells, hepatocytes, macrophages, and retinal pigmented epithelial cells on surfaces grafted with PIPAAm temperature-responsive polymer [4—7]. Under optimal conditions, we found no differences between the PIPAAm-grafted and ungrafted TCPS dishes on attachment, spreading, growth, confluent cell density, and cell morphology at 37°C. Stress fibers, peripheral bands, as well as focal contacts were also established in the same way as shown in Fig. 1 [8]. However, cells already spread on the PIPAAm-grafted surfaces at 37°C were prompted to detach simply by reducing the medium temperature without any enzymatic digestion or divalent cation chelators. Optimal conditions that allow cell detachment depend on cultured cell type. Generally, adhesive cells require higher density of grafted PIPAAm for cell detachment after reducing culture temperature. This cell detachment is inhibited by various chemical inhibitors of cell functions, such as sodium azide for ATP synthesis, genistein for protein tyrosine

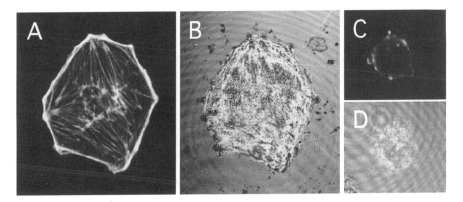

Fig. 1. Cell adhesion and detachment on temperature-responsive culture surfaces. Bovine aortic endothelial cells were plated on PIPAAm-grafted culture dishes. Cells spread and developed stress fibers during the culture at 37°C (**A**). Interference reflection microscopy revealed that the cells also established focal contacts (**B**). Then, culture temperature was decreased to 20°C. The cells spontaneously detached from the surfaces. In the detaching spherical cells stress fibers (**C**) and focal contacts (**D**) disappeared.

phosphorylation, phalloidin and cytochalasin for F-actin dynamics [6,8]. This finding suggests a requirement of cellular structural and metabolic activity in the detachment process.

Cells recovered by low-temperature treatment consistently maintain their differentiated functions more highly than cells recovered by trypsinization, because membrane proteins susceptible to destruction by usual enzymatic digestion are preserved intact. Highly trypsin-susceptible cells such as aortic endothelial cells, hepatocytes [5], and retinal pigmented epithelial cells [7] were examined. Low-temperature treatment did not alter endogenous prostacyclin (PGI_2) secretion by endothelial cells, albumin secretion by hepatocytes, and retinal profiles of retinal pigmented epithelial cells. In this sense, nonenzymatic cell harvest from the temperature-responsive culture surfaces is noninvasive, gentle, and harmless to cells. In fact, we found that these intelligent culture surfaces are very useful for primary culture. Common procedures to dissociate cells from tissues in primary culture use enzymatic digestion. Primary cells isolated using only highly purified collagenase, that does not disturb noncollagenous proteins, retain relatively higher functions. This observation suggests a possibility that general proteolytic treatment injures cells irreversibly by membrane protein degradation. At least in part, this enzymatic handling is responsible for why secondary cultured cells gradually lose their specific functions after repeated subcultivation and passages using trypsin.

We have recently succeeded in establishing blood-organ barrier cell lines derived from transgenic animals harboring the temperature-sensitive SV40 large T-antigen gene [9]. Trypsin-free subculture with our temperature-responsive culture surfaces was utilized for all steps. Immortalized cell lines of brain capillary endothelial cells and astrocytes express characteristic proteins such as the glucose transporter, GLUT1, and glial fibrillary acidic protein, GFAP, respectively. Confluent, viable recovered cell sheets of the cultured lines are currently applied for studies on drug delivery to the brain.

Another example involves microglial cells. Small glial cells are derived from monocytes and are capable of enlarging to exhibit macrophage-like phenotypes after activation by external stimuli. Microglial cells are extremely sticky and difficult to culture by conventional means. Only 50% of these cultured cells are recovered from commercial TCPS dish surfaces using trypsin. Significantly, this same trypsin treatment activates these cells. Activated microglial cells lose their phenotypes found in normal brain tissue and enlarge. However, by simply reducing cell culture temperature, nearly all cultured microglial cells are recovered from grafted temperature-responsive culture surfaces. Microglial cells recovered without enzymatic digestion maintain their normal phenotypes.

Two-dimensional manipulation of large intact cultured cell sheets

When culture temperature is decreased after cells proliferate to confluency, cells spontaneously detach from the PIPAAm-grafted surfaces maintaining intact cell-cell junctions that are otherwise highly susceptible to enzymatic digestion or chelation of divalent cations (Fig. 2). Cytoskeleton and junctional complexes remained organized even in cells detaching in confluent monolayer sheets. Immunofluorescence microscopy and immunoblotting with anti-fibronectin antibodies revealed that extracellular matrix (ECM) fibronectin deposits and accumulates primarily between the basal side of the cells and either the PIPAAm-grafted surfaces or ungrafted culture dish surfaces. Although trypsin digestion completely destroyed the deposited fibronectin matrix, this important matrix retained intact when recovered by lowering culture temperature [10]. This was also confirmed by transmission electron microscopy. No ECM remnants are observed on the PIPAAm-grafted surfaces after cell recovery. Intact, viable

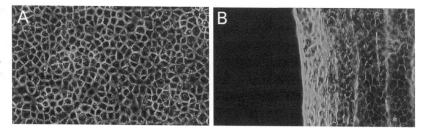

Fig. 2. Phase contrast microscopy of detaching endothelial cell monolayers from temperature-responsive culture dishes. Bovine aortic endothelial cells were plated on culture dishes grafted with PIPAAm, and cultured for 1 week to reach confluency at 37°C (**A**). Then, medium temperature was reduced to 20°C. Confluent cell sheets were detaching from culture surfaces from the cell sheet periphery (**B**). Note that endothelial cells retain cell-cell junctions.

monolayer cell sheets are recovered together with ECM deposited during cell culture. Fibronectin matrix adherent to the basal side of the harvested cell sheets can function as a natural substance to attach cell sheets onto other surfaces. In fact, cell sheets recovered from temperature-responsive surfaces easily adhere onto other surfaces.

Cell sheet transplantation would be a promising general technique in tissue engineering, extending capabilities already achieved with cultured keratinocyte sheets [11]. Cultured keratinocytes are now widely utilized in clinical use as an artificial skin, although in this case, cell sheets are recovered from culture dishes by enzymatic digestion. Since differentiated keratinocytes show multilayer cell stratification, the thickness and rigidity of keratinocyte sheets are exceptionally high. This is the reason why enzymatic digestion is applicable to the recovery of keratinocyte sheets. It is precisely this reason, however, that enzymatic digestion is inapplicable to recover thin and fragile monolayer sheets of many other cell types. Enzymatic digestion often dissociates monolayer cell sheets into isolated single cells. Confluent monolayer cell sheets recovered from temperature-responsive surfaces simply by low-temperature treatment represent a significant advance with notable potential.

We have further developed methods to transfer harvested monolayer cell sheets from temperature-responsive surfaces to other culture dishes (Fig. 3) [12]. Cultured cell sheets are inherently very fragile and apt to aggregate and shrink during the detachment process: support of harvested cell sheets is required to prevent deformation and shrinkage during transfer. For this purpose, we have developed two different cell culture transfer methods represented in Fig. 3. The first method utilizes chitin membranes or cover glass slips as supports for harvested cell sheets during cell sheet transfer. In this method, basal sides of harvested cells attach to new culture substrates after the transfer, resulting in unchanged cell topology in terms of apical-basal polarity. The second method shown in Fig. 3 uses commercial Cell Culture Inserts^{TM} (Falcon) for transfer. The bottom surface of Cell Culture Inserts is coated with an adhesive matrix, such as type I collagen, and utilized both as a transfer support as well as a new substrate. In this case, formerly basal sides of harvested cells are exposed to the culture medium after sheet transfer, resulting in complete reversal of apical-basal cell topology on new culture surfaces. In both cases, confluent cell sheets were recovered from PIPAAm-grafted culture dishes and transferred intact onto other culture dishes. In this manipulation without EDTA or trypsin, no disruption of cell-cell junctions is observed by light and electron microscopy. Characteristic cell morphology remains the same as that before cell transfer. Recovered cells retain approximately 90% viability as well as their original phenotypic function as judged from tissue-type plasminogen activator secretion in endothelial cells. Using this technique we were able to confirm the apical-basal distribution and functions of proteins expressed on cell plasma membranes, such as GLUT1 and neutral amino acid transporter in vitro for the first time [9].

Surprisingly, we discovered that hepatocyte functions were upregulated during

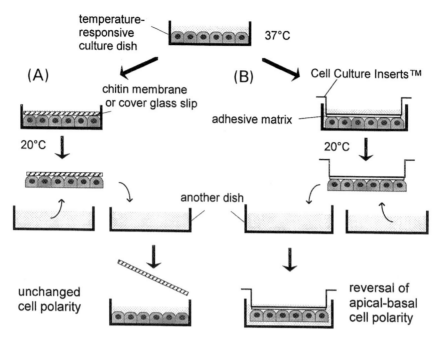

Fig. 3. Two-dimensional cell sheet manipulation. Two different transfer methods of confluent mono-layer cell sheets were developed. **A:** Chitin membranes or cover glass slips are utilized as supports to prevent shrinkage of cell sheets during transfer. Basal sides of cells attach to new culture substrates after the transfer, resulting in unchanged cell topology in terms of apical-basal polarity. **B:** The bottom surface of Cell Culture Inserts[TM] (Falcon) is coated with type I collagen and utilized as a transfer support as well as a new substrate. In this case, formerly basal sides of cells are exposed to the culture medium after the transfer, resulting in complete reversal of cell topology on culture surfaces.

this transfer. As on ungrafted TCPS dishes coated with or without adhesive proteins, hepatocytes on temperature-responsive surfaces rapidly lose their characteristic cuboidal shape: they spread and flatten. Accompanying this loss of cuboidal shape, hepatocytes also lose their differentiated cell functions which are seen in vivo. Decreases in culture temperature result in rapid hydration of grafted PIPAAm chains, and weakening of the cell-surface interactions, allowing detachment of hepatocyte sheets from these surfaces. Detached hepatocytes in monolayer sheets retain their cell-cell junctions and cytoskeleton that exert contractile forces to shrink cells. Therefore, after surface detachment, cell contraction occurs and harvested hepatocyte sheets shrank to 70% of the original total cell area seen prior to transfer. Cell contraction also resulted in an increase in cell height and recovery of the characteristic cuboidal shape. Recovered sheets were transferred onto type I collagen-covered cell inserts and cultured for another 24 h. After transfer, albumin secretion and ornithine carbamoyltransferase activity were increased to 300 and 166% of untransferred controls, respectively. These results imply that temperature-responsive surfaces can also provide a novel way to recover native cellular functions after cell growth for utilization in tissue engineering. Cell sheet transplantation experiments using temperature-responsive surfaces are now ongoing with retinal pigmented epithelial cells [7] and hepatocytes [13].

Stratified cell sheet culture

Using the above-mentioned cell sheet recovery-transfer technique, we have succeeded in fabricating stratified cell sheets to reconstruct tissue-like structures (Fig. 4) [13]. In natural organs, the parenchyma comprises intimately associated cell sheets. For example, liver comprises sheets of hepatocytes and endothelial cells that are interconnected to form a continuous three-dimensional tissue lattice. In liver lobules, radially disposed hepatocyte sheets are covered with an overlayer of endothelial cells on the side adjacent to the sinusoid. In order to develop a novel hybrid artificial organ with a three-dimensional structure similar to normal tissues, we have focused on the reconstruction of cell sheets, not isolated co-cultured single cells, into tissue-like structures as a breakthrough technology. To accomplish this, hepatocytes and endothelial cells were cultured separately on TCPS dishes grafted with PIPAAm. Initially, a sheet of confluent hepatocyte monolayer was recovered with a Cell Culture Insert, allowing reversal of apical-basal sheet orientation. This cell sheet was then placed intact onto the apical surface of a confluent endothelial cell monolayer adhered to a PIPAAm-grafted sur-

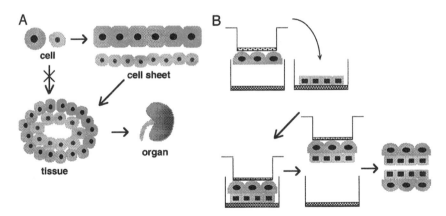

Fig. 4. Schematic drawing of reconstruction of tissue architectures using two- and three-dimensional cell sheet manipulation techniques. In order to develop a novel hybrid artificial organ with a three-dimensional structure similar to normal tissues, we have focused on the reconstruction of cell sheets, not isolated co-cultured single cells, into tissue-like structures (**A**). For example, we reconstructed liver-lobule-like structure from hepatocyte and endothelial cell sheets (**B**). Hepatocytes and endothelial cells are cultured on separate culture dishes grafted with PIPAAm. Initially, confluent hepatocyte monolayer sheets is recovered with using a Cell Culture Insert. Then, this cell sheet is transferred intact onto the apical surface of a confluent endothelial cell monolayer adhered to PIPAAm-grafted surface, and subjected to low temperature to obtain intact stratified double layers of cell sheets. The stratified two cell layers are combined and transferred to another dish to continue the culture.

106

face. The two cell layers adhere readily and rapidly to each other after a short time of coincubation, likely due to the intact ECM recovered and codeposited together with the overlaying hepatocyte sheet. This stratified system was then subjected to low temperature to obtain the intact stratified double-layer of viable cell sheets, and transferred to another well. This stratified cell sheet culture highly resembles liver lobules at least histologically. Transmission electron microscopy revealed that bile canaliculi-like structures were established between hepatocytes, and Disse's space-like structures filled with thin ECM were also observed between hepatocytes and endothelial cells. Cultured overlaid lamellar cell sheets preserve each cell phenotype and basic cell functions. Further biochemical analyses of hepatocyte and endothelial cell functions in the stratified double-layer co-culture are now ongoing.

Two- and three-dimensional manipulations of cell sheets using culture dishes grafted with a temperature-responsive polymer should prove useful as a fundamental, generalized technique in tissue engineering. Many new experiments useful for fundamental studies of basic cell biology, cell-cell interactions, and structural matrix are now possible. Applications in a number of tissue regeneration areas hold promise.

Acknowledgements

We would like to acknowledge Prof D.W. Grainger for critical reading of the manuscript as well as insightful discussions. We are also thankful to Ms A. Kushida and Ms C. Konno for their great help in preparation of this maniscript. The present work was supported by the Japan Society for the Promotion of Science, "Research for the Future" Program (JSPS-RFTF96I00201). Interactions with Prof Grainger were made possible by an NSF-JSPS Short-Term Invited Professor Award.

References

1. Yamada N, Okano T, Sakai H, Karikusa F, Sawasaki Y, Sakurai Y. Thermo-responsive polymeric surfaces; control of attachment and detachment of cultured cells. Makromol Chem Rapid Commun 1990;11:571–576.
2. Bae YH, Okano T, Kim SW. Temperature dependence of swelling of crosslinked poly(N,N'-alkyl substituted acrylamide) in water. J Polym Sci Polym Phys 1990;28:923–936.
3. Heskins M, Guillent JE, James E. Solution properties of poly (N-isopropylacrylamide). J Macromol Sci Chem 1968;A2:1441–1455.
4. von Recum HA, Kim SW, Kikuchi A, Okuhara M, Sakurai Y, Okano T. Retinal pigmented epithelium culture on thermally responsive polymer porous substrates. J Biomater Sci Polymer Edn 1998;9:1241–1254.
5. Okano T, Yamada N, Sakai H, Sakurai Y. A novel recovery system for cultured cells using plasma-treated polystyrene dishes grafted with poly (N-isopropylacrylamide). J Biomed Mater Res 1993;27:1243–1251.
6. Okano T, Yamada N, Okuhara M, Sakai H, Sakurai Y. Mechanism of cell detachment from temperature-modulated, hydrophilic-hydrophobic polymer surfaces. Biomaterials 1995;16:

297–303.

7. von Recum HA, Kim SW, Kikuchi A, Okuhara M, Sakurai Y, Okano T. Novel thermally reversible hydrogel as detachable cell culture substrate. J Biomed Mater Res 1998;40:631–639.

8. Yamato M, Okuhara M, Karikusa F, Kikuchi A, Sakurai Y, Okano T. Signal transduction and cytoskeletal reorganization are required for cell detachment from cell culture surfaces grafted with a temperature-responsive polymer. J Biomed Mater Res 1999;44:44–52.

9. Tetsuka K, Nagase K, Saeki S, Tomi M, Hosoya K, Yanai N, Obinata M, Ueda M, Kikuchi A, Okano T, Terasaki T. Blood-organ barrier cell lines derived from transgenic animals harboring temperature-sensitive SV40 large T-antigen gene: establishment and characterization of brain capillary endothelial and astrocyte cell lines. In: Conference on Challenges for Drug Delivery and Pharmaceutical Technology. Tokyo, Japan, 1998;93.

10. Kushida A, Yamato M, Konno C, Kikuchi A, Sakurai Y, Okano T. Decrease in culture temperature releases monolayer endothelial cell sheets together with deposited fibronectin matrix from temperature-responsive culture surfaces. J Biomed Mater Res (In press).

11. Gallico GG, O'Connor NE, Compton CC, Remensnyder JP, Kehinde O, Green H. Permanent coverage of large burn wounds in autologous cultured human epithelium. N Engl J Med 1984; 311:448–451.

12. Kikuchi A, Okuhara M, Karikusa F, Sakurai Y, Okano T. Two-dimensional manipulation of confluently cultured vascular endothelial cells using temperature-responsive poly(N-isopropyl-acrylamide)-grafted surfaces. J Biomater Sci Polym Edn 1998;9:1331–1348.

13. Yamato M, Okuhara M, Karikusa F, Kikuchi A, Sakurai Y, Okano T. Reconstruction of liver lobule-like structure with cell sheets of hepatocytes and endothelial cells recovered from culture dishes grafted with temperature-responsive polymer. Trans Soc Biomater 1998;XXI:82.

JSPS TISSUE ENGINEERING PROJECT: BIOMATERIALS FOR TISSUE ENGINEERING

Reconstituted collagen assemblies as building blocks for the construction of multicellular system in vitro

Toshihiko Hayashi[1], Motohiro Hirose[1], Masayuki Yamato[2], Kazunori Mizuno[1], Koichi Nakazato[1], Eijiro Adachi[3], Hiroaki Kosugi[1], Yasuhiro Sumida[1], Teruo Okano[2], Kiwamu Yoshikawa[1], Yasushi Takeda[1], Seiichiro Takahashi[1] and Yasutada Imamura[1]

[1]Department of Life Sciences, Graduate School of Arts and Sciences, The University of Tokyo; [2]Institute of Biomedical Engineering, Tokyo Women's Medical University, Tokyo, Japan; and [3]Kitasato University Medical School, Kanagawa, Japan

Abstract. We have proposed graded filamentous aggregates of different types of collagen as a general tissue skeletal structure [1]. Type IV collagen essentially forms polygonal meshwork, while type V collagen forms thin collagen fibrils with a banding pattern similar to, if not identical with, type I collagen fibrils. Type I collagen is a major component of the collagen fibrils in all connective tissues except for hyaline cartilage. The findings that various functions of fibroblasts were affected by surrounding type I collagen fibrils in culture have tempted us to examine whether this can be extended further to a more general situation. Thus, the question was addressed whether cell growth, differentiation or migration of other types of cells are affected by surrounding consolidated collagen aggregates composed of different types of collagen. We have succeeded in developing the reconstituted type IV collagen and type V collagen aggregates that can be used as cell culture substrates. Then we focused on pericytes or smooth-muscle-related cells, as well as endothelial cells that may be situated close to these aggregates in vivo. In comparison with the previously used substrates including plastic dish, the dish coated with monomeric type I collagen, type I collagen gel, and Matrigel, on the type IV collagen gel the smooth-muscle-related cells showed a unique morphology with an extremely elongated shape, eventually forming cell-to-cell junctions. The whole cell population was connected to each other through cell processes, ultimately forming a multicellular network. The multicellular network suggests a typical contractile state of the smooth muscle cells. This is the first report that the extracellular consolidated environment may regulate smooth muscle cell differentiation into the contractile state in the presence of bovine fetal serum. Furthermore, repression of growth and induction of cell-to-cell junction on the type IV collagen gel were also found for the myofibroblast-like cells that had acquired proliferating and synthetic activities through many passages. The results suggest that culture on the type IV collagen gel may possibly provide a new method for obtaining the differentiated smooth-muscle-related cells, even once they have acquired the synthetic property. Type V collagen fibrils showed as strong an inhibition of endothelial cell proliferation as type I collagen fibrils. The effect of type V collagen appeared to depend greatly on whether serum was present or not. Different types of collagen may have cell- or tissue-specific influences on cell growth and differentiation. Thus, consolidated collagen aggregates of type IV collagen and type V collagen, together with type I collagen, might be used with an appropriate combination of different types of cells for forming functional multicellular systems in vitro or for regeneration of functional organs in vivo.

Address for correspondence: Toshihiko Hayashi, Department of Life Sciences, Graduate School of Arts and Sciences, The University of Tokyo, 3-8-1 Komaba, Meguro-ku, Tokyo 153-8902, Japan.

Keywords: cell differentiation, smooth muscle cell, type V collagen fibrils, type I collagen gel, type IV collagen gel.

Introduction

The characteristic feature of cells in a multicellular system (e.g., organ or tissue) is control. Cells must not divide uncontrollably. They must stay within their tissues rather than wander off to colonize new locations in the body. The mechanisms by which cell division and migration are controlled are gradually becoming understood at the molecular level (modified from [2]). The extracellular matrix is one of the essential components of the multicellular system in vivo. Molecular structures of extracellular matrix components, and interactions of the components with each other or with cells have been intensively elucidated. Supramolecular aggregates, such as type I collagen fibrils, rather than molecules, could be existing forms in vivo, and thus functional forms of the extracellular matrix components including collagenous proteins.

The collagenous protein family represents one of the most abundant proteins of large vertebrates such as the human. It has been known that collagen supramolecular structures provide scaffoldings of matrix architecture of various organs. Thus, deficiency of type I collagen may cause tissue fragility in organs, such as bone, while excessive deposition of type I collagen may be taken as pathogenesis of some diseases, typically organ fibroses.

The matrix architecture is, in principle, constructed through specific interactions of extracellular matrix components secreted from the cells as well as self-assembly. Furthermore, reiterative interactions between collagen aggregates and connective tissue cells may modify arrangements of supramolecular aggregates [3]. This idea is based on the finding that fibroblasts or other mesenchymal cells, including smooth muscle cells, contract reconstituted type I collagen gel when they were cultured within the gel [3]. These observations suggest strong interactions between cells and consolidated extracellular collagen aggregates at least in the case of type I collagen and fibroblasts.

It has not been examined whether supramolecular assemblies of other types of collagen influence the surrounding cells. In this report, we will propose the hypothesis that specific collagen supramolecular aggregates as extracellular, consolidated environments regulate cellular functions including cell growth, cell shape, cell migration, cell adhesions including cell-to-cell contact, and gene expression. We will introduce the possibility that the reconstituted type IV and type V collagen supramolecular assemblies serve as new biomaterials for construction of multicellular system in vitro.

Materials and Methods

As cell culture substrates, two yet unexamined collagen aggregates were tested: reconstituted type IV collagen aggregates in the form of polygonal meshwork (macroscopically a form of gel [4]); and reconstituted type V collagen aggregates in the form of ribbon-like thin fibrils [5] (Fig. 1).

Three different types of cells were examined: vascular endothelial cells,

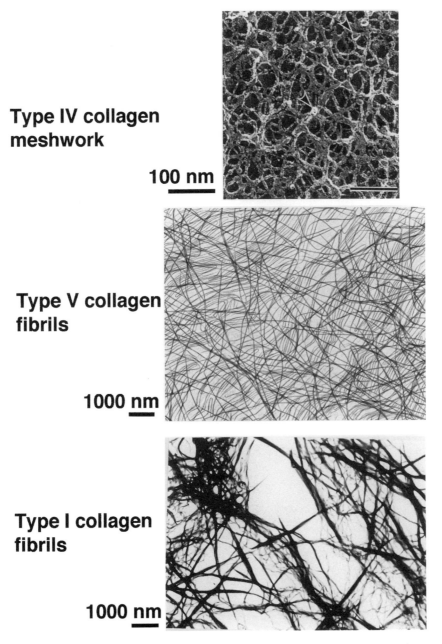

Type IV collagen meshwork

100 nm

Type V collagen fibrils

1000 nm

Type I collagen fibrils

1000 nm

Fig. 1. Comparison of the shape and size in the electron micrographs of reconstituted aggregates of type I collagen, type IV collagen and type V collagen.

112

smooth-muscle-related cells (vascular smooth muscle cells, kidney mesangial cells, hepatic stellate cells), and fibroblasts. Myofibroblast-like cells were obtained from the primary rat hepatic stellate cells by in vitro repeated passages up to 28 PDL (population doubling level). They were also examined in order to see whether the cells that once had acquired synthetic stages could be reversed to the differentiated state, in view of the possible involvement of myofibroblasts in interfering with the regeneration of functional tissues or organs from a fibrotic or cirrhotic state.

Results

Isolated type I collagen, type IV collagen and type V collagen can form, under selected conditions, the specific supramolecular assemblies in the electron micro-scopic observations (Fig. 1): type I collagen forms fibrils of varying diameters with branching (macroscopically a gel form); type IV collagen forms polygonal meshwork with pores of on average ~ 18 nm in diameter (macroscopically fragile gels) [3]; and type V collagen forms thin-width tape-like fibrils without branch-ing. The apparent similarities of the reconstituted aggregate structures to the in vivo supramolecular structures, as well as the tissue localization of different types of collagen, have allowed us to hypothesize that the major four collagen types may comprise the scaffoldings of various tissues as shown in Fig. 2 [1]. The arrangement of the collagenous protein aggregates, we further hypothesize, may accommodate differentiated cells in a graded manner corresponding to endothe-lial cells, smooth muscle cells and fibroblasts in the case of blood vessel wall (Fig. 2).

Some results on the proliferation of cells cultured on type I collagen gel and type IV collagen gel are summarized in Table 1 [3]. Endothelial cells were quies-cent on the type IV collagen polygonal meshwork. In contrary to previous reports, we found that the type V collagen, especially in the form of fibrils, was

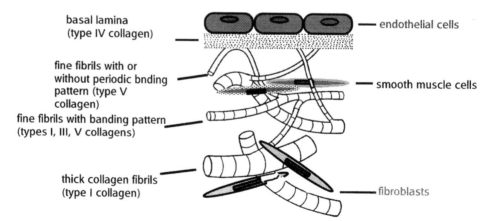

Fig. 2. Schematic vessel model representing the relationship of cell types and collagen types.

Table 1. Proliferation of three different cell types cultured on type IV collagen, type I collagen, and type V collagen in different forms.

	Endothelial	Smooth muscle cells	Fibroblasts
Type IV collagen			
Molecules	?	Proliferative	Proliferative
Rigid gel	No	No	Proliferative
Type I collagen			
Gel	No	Proliferative	Retarded
Densely packed fibrils	?	No	No
Type V collagen			
Fibrils	Little	?	?

?: not examined fully or inconclusive.

not poor in adhering to the vascular endothelial cells. The initiation of endothelial growth or the growth rate was markedly repressed in the presence of 10% bovine fetal serum in comparison with the plastic dish coated with type I collagen molecules. Growth of the endothelial cells from bovine artery was more strongly inhibited by the culture on the reconstituted type I collagen gel or on the reconstituted type V collagen fibrils than on the plastic dish coated with type I collagen molecules.

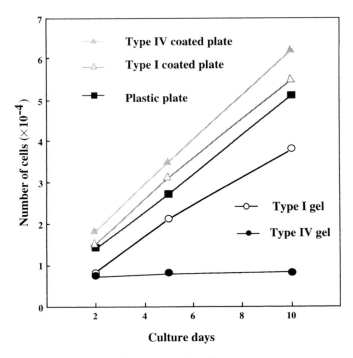

Fig. 3. Growth curves of smooth muscle cells.

Smooth muscle cells are quiescent on the type IV collagen gel (Fig. 3). Mesangial cells from kidney glomerulus and hepatic stellate cells were also arrested on the type IV collagen gel. Growth inhibition of the smooth-muscle-related cells on the type IV collagen gel is not due to the type IV collagen protein structure, since they grew rather rapidly on the dish coated with type IV collagen molecules. In fact, most rapid growth was reproducibly observed on the monodispersed type IV collagen. All these smooth muscle-related cells, however, grew actively when they were cultured on the type I collagen gel, as shown in Fig. 3, suggesting that the growth inhibition on the type IV collagen gel is not solely due to the physical state of the substrate in a gel form. The smooth muscle cells, as well as the fibroblasts, are known to contract the type I collagen gel in which the cells were cultured. They appeared to become quiescent even in the presence of 10% serum in the contracted collagen gel.

Fibroblasts only grew slowly in the collagen type I gel and stopped growing after gel contraction without mechanical stresses. However, fibroblasts rapidly proliferated on the type IV collagen.

Myofibroblast-like cells that had been obtained through many passages of rat primary hepatic stellate cells showed very similar behavior common with smooth-muscle-related cells. They grew even more rapidly on type I collagen gel than on the primary stellate cells. However, the myofibroblast-like cells did not proliferate at all on the type IV collagen gel. The type IV collagen gel is the first example that the total repression of myofibroblast-like cells was caused by the extracellular environments in the presence of 10% fetal bovine serum and/or other growth factors.

The cell shape was most dramatically affected by the type IV collagen gel in the case of smooth-muscle-related cells including myofibroblast-like cells, in which the cells were retarded in spreading and elongating [6], but yet highly elongated in shape to get in touch with the elongated tips of neighboring cells, resulting in the formation of cell-to-cell junctions (Fig. 4). The whole cells in culture eventually formed a single multicellular network. The multicellular network collapsed to give rise to many rounded cells upon treatment with bacterial collagenase that only degrades the collagenous polypeptide sequences.

It is quite intriguing that the smooth muscle cells most rapidly proliferated on the dish coated with nonaggregated type IV collagen. The results may suggest that the control of smooth muscle cell proliferation is dependent crucially on the physical property of the type IV collagen, and independent of the chemical structure including the triple-helical conformation.

We do not know what types of cell junctions are formed at the molecular or electron microscopic level. Apparent facilitation of cell junction formation by type IV collagen gel might involve the signaling of the cell-substrate interaction to the formation of cell-to-cell contact and junction. The fibroblasts become quiescent in the type I collagen gel after gel contraction, but they proliferated on the type IV collagen gel as much as they did on the type I collagen gel.

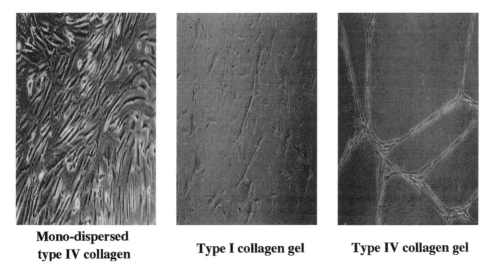

Mono-dispersed type IV collagen   **Type I collagen gel**   **Type IV collagen gel**

Fig. 4. Differential effects of type I collagen gel and type IV collagen gel or monodispersed type IV collagen on the cell shape of smooth muscle cells.

Discussion

The results have tempted us to speculate that the tissue differentiation might be maintained by "appropriate" interactions between the differentiated cells and matrix environments, particularly collagen supramolecular aggregates to which cells adhere. In order to test the above hypothesis, we currently plan to construct a multilayer organ-like multicellular system, as shown in Table 2. The three layers may be composed of endothelial cells-type IV collagen gel-liver stellate cells-type V collagen/ type III collagen-type IV collagen-parenchymal cells for the liver sinusoid model. In the case of small blood vessels, the construct may be endothelial cells-type IV collagen-type V collagen-smooth muscle cells-type V collagen-type III collagen-type I collagen-fibroblasts.

The examination of collagen-type distribution in different tissues and ultra-

Table 2. Three layers of differentiated tissue-like constructs.

Gradient of cell types	Corresponding supramolecular assemblies reconstituted from isolated collagen
Endothelial cells/ epithelial cells	Polygonal meshwork of type IV collagen
Pericytes/ smooth muscle cells	Type IV collagen gel, fine fibrils of type V collagen and type I collagen gel
Fibroblasts	Condensed type I collagen fibrils

structural characteristics of filamentous aggregates in conjunction with the differentiated cells accommodated in the extracellular matrices has led us to hypothesize that a gradient arrangement of the skeletal structure composed of different types of collagen might be a general architectural scheme in relation to a graded distribution of differentiated cells [1].

Of the effects of type IV collagen on cell functions in culture, the gel form is important in inhibiting the growth of smooth muscle cells, mesangial cells, and hepatic stellate cells, all of which are said to be involved in the fibrotic pathogenesis; arteriosclerosis, glomerular sclerosis, and liver cirrhosis. The reason why only type IV collagen in the gel form is effective among the substrates examined remains to be elucidated. The NC1 domain, particularly in the conformation changed upon its binding to the type IV collagen triple-helical domain, might be related to its unique effectiveness.

A differential effect of type V collagen fibrils from type I collagen fibrils on the endothelial growth in culture [5,7] may be biologically intriguing, because in tissues the vascular endothelial cells in blood vessels are closer to type V collagen fibrils than type I collagen. Sprouting of blood vessels in normal tissues might be controlled by the type V collagen fibrils that may interfere with endothelial cell proliferation under serum-depleted conditions, while under serum-supplemented conditions the endothelial cells might adhere to the type V collagen fibrils and grow. Characteristics of the type V collagen fibrils in comparison with the type I collagen fibrils include a thinner diameter, strong heparin-binding property, or no branching or nongel form.

Type IV collagen is distributed only at specific sites of normal liver sinusoids, but in fibrotic or cirrhotic tissues, the distribution of the type IV collagen is scattered all over the tissues [8]. The disorganized distribution of collagenous proteins is thought to be due to abnormal proliferation of smooth-muscle-related cells that may produce excessive collagenous proteins such as type I collagen, type III collagen, and type V collagen.

It is thought that these smooth-muscle-related cells are responsible for organ fibrosis or cirrhosis, where the cells derived from smooth-muscle-related cells that have acquired highly proliferating potential also produce excessive collagenous fibrils. The control mechanism of these cells whether they are synthetic and proliferating, or contractile and nonproliferating is not known. The effect of the type IV collagen gel found in in vitro culture in the present study might provide a new possibility that adjustment of the extracellular environment could be a new way to repress synthetic and proliferating states of these myofibroblast-like cells.

Regulation of angiogenesis is now being focused as a new therapy for tumor regression. Degradation of the extracellular matrix with MMP (matrix metalloproteinases) or proteases secreted by tumor cells might initiate tumor angiogenesis. Thus, protease inhibitors against such proteases are proposed as new therapeutic drugs for tumor regression. In turn, potential inhibitors of vascular endothelial cell proliferation, endostatin for example, are reported to be potent for tumor regression by suppressing tumor angiogenesis. In any case, if the scaffolding of the blood vessels of tumor tissues is different from that of the normal vessel, targeting the tumor angiogenesis could be realized with a specific method that recognizes only the tumor-specific structure. The multicellular system composed of the organized collagen supramolecular architecture together with the mesenchymal cells such as smooth muscle cells might be a useful system to evaluate the effectiveness of potential angiogenesis inhibitors.

Local optimization of the interactions between cells as well as the interactions between cells and extracellular signals including extracellular matrix environments might be important in the formation of organ structure that is biologically important for fulfilling organ functions. Reiterative interactions between consolidated aggregates (type I collagen fibrils, type IV collagen polygonal meshwork, and type V collagen fibrils) and cells (endothelial cells/epithelial cells, smooth muscle cells, and fibroblasts) are to be taken into consideration for the local optimization.

We conclude that the collagenous proteins, assembled structures in particular, serve as functional as well as structural building blocks required for constructing and maintaining the multicellular system.

Acknowledgements

The present study was supported in part by the Scientific Research Grant from the Ministry of Education, Culture, Sports and Sciences of Japan: Grant-in-Aid for Scientific Research on Priority Areas (09229219 Functionally Graded Materials, 09217210 Supramolecular Structure), Grant-in-Aid for Developmental Scientific Research (07558249), by The Japan Society for the Promotion of Science, "Research for the Future" Program (JSPS-RFTF96I00201) and by the Program for Promotion of Fundamental Studies in Health Science of The Organization for Pharmaceutical Safety and Research (OPSR).

References

1a. Adachi E, Hopkinson I, Hayashi T. Basement-membrane stromal relationships: interactions between collagen fibrils and the lamina densa. Int Rev Cytol 1997;173:73—156.

1b. Adachi E, Hayashi T. Anchoring of epithelia to underlying connective tissue: evidence of frayed ends of collagen fibrils directly merging with meshwork of lamina densa. J Electron Microsc 1994;43:264—271.

2. Wood EJ, Smith CA, Pickering WR. Life Chemistry and Molecular Biology. London: Portland Press Ltd., 1997;216.

3a. Nakazato K, Muraoka M, Adachi E, Hayashi T. Gelation of the lens capsule type IV collagen solution at a neutral pH. J Biochem 1996;120:889—894.

3b. Adachi E, Takeda Y, Nakazato K, Muraoka M, Iwata M, Sasaki T, Imamura Y, Hopkinson I, Hayashi T. Isolated collagen IV retains the potential to form an 18-nm polygonal meshwork of the lamina densa. J Electron Microsc 1997;46:233—241.

4a. Nishiyama T, Tsunenaga M, Akutsu N, Horii I, Nakayama Y, Adachi E, Yamato M, Hayashi T. Dissociation of actin microfilament organization from acquisition and maintenance of elon-

118

gated shape of human dermal fibroblasts in three-dimensional collagen gel. Matrix 1993; 13:447—455.

4b. Yamato M, Adachi E, Yamamoto K, Hayashi T. Condensation of collagen fibrils to the direct vicinity of fibroblasts as a cause of gel contraction. J Biochem 1995;117:940—946.

4c. Yamato M, Hayashi T. Topological distribution of collagen binding sites on the fibroblasts cultured within collagen gels. In: Ninomiya Y, Olsen BJ, Ooyama T (eds) Extracellular Matrix-Cell Interaction Molecules to Diseases. Tokyo: Japan Scientific Societies Press and S. Karger AG, 1998;123—140.

4d. Hayashi T, Yamato M, Adachi E, Yamamoto K. Unique features of type I collagen as a regulator for fibroblast functions: iterative interactions between reconstituted type I collagen fibrils and fibroblasts. In: Tsuruta T, Doyama M, Seno M, Imanishi Y (eds) New Functionality Materials, vol B. Synthesis and Functional Control of Biofunctionality Materials. Amsterdam: Elsevier Science Publishers BV, 1993;239—246.

5. Mizuno K, Hayashi T. Separation of the subtypes of type V collagen molecules, $[\alpha 1(V)]_2 \alpha 2(V)$ and $\alpha 1(V) \alpha 2(V) \alpha 3(V)$, by chain composition-dependent affinity for heparin; single $\alpha 1(V)$ chain shows an intermediate heparin affinity of the type V collagen subtypes composed of $[\alpha 1(V)]_2 \alpha 2(V)$ and of $\alpha 1(V) \alpha 2(V) \alpha 3(V)$. J Biochem 1996;120:934—939.

6. Hirose M, Nakazato K, Mizuno K, Adachi E, Imamura Y, Hayashi T. Structure and function of type IV collagen and type V collagen. In: Akaike T, Okano T, Akashi M, Terano M, Yui N (eds) Advances in Polymeric Biomaterials Science. Tokyo: CMC Co. LTD., 1997;153—166.

7. Yamato M, Sumida Y, Mizuno K, Adachi E, Sakurai Y, Okano T, Hayashi T. Strong inhibitory effect of type V collagen fibrils on bovine aortic endothelial cell growth. Abstracts of the Third Congress of the Asian-Pan-Pacific Organization for Cell Biology, 24—28 Aug, 1998.

8. Yoshida T, Matsubara O, Adachi E, Kino J, Asamatsu C, Takeda Y, Hayashi T. Increased deposition and altered distribution of basement membrane-related collagens in human fibrotic or cirrhotic liver. Connect Tis 1997;29:189—198.

Tissue Engineering for Therapeutic Use 3.
Y. Ikada and T. Okano, editors.

Biocompatible alginate scaffolds enabling prolonged hepatocyte functions in culture

R. Glicklis[1], S. Zmora[1], L. Shapiro[1], S. Cohen[1], R. Agbaria[2] and J.C. Merchuk[1,3]

[1]Unit of Biotechnology, [2]Soroka Medical Center, and [3]Department of Chemical Engineering, Ben-Gurion University of the Negev, Beer-Sheva, Israel

Abstract. Temporary hepatic support utilizing hepatocytes can save the life of patients suffering from liver failure since it can bridge the time needed for donor-patient matching in liver transplantation, and can in some cases give the time for liver regeneration without recurring to transplantation. Tissue engineering is one approach to achieve such supports. In this approach polymeric scaffolds play a critical role in maintaining the viability and functions of the hepatocytes.

In this paper, a novel type of scaffold for hepatocyte culture and transplantation is reviewed. The scaffold consists of a sponge fabricated from alginates by a multistep technique that includes gelation, freezing and lyophilization. It has an appropriate porosity and pore size to accommodate a large cell mass and to allow blood vessel ingrowth after implantation; the pores are interconnected enabling the organization of the cells into a tissue; it has appropriate mechanical properties to assure flexibility and structure conservation; the matrix is bioresorbable and in addition is hydrophilic, thus facilitating cell seeding and distribution within the matrix. The hepatocytes seeded within the sponges are distributed homogeneously and are located mostly within the pores.

The cells aggregate within the sponges until reaching a stationary size matching that of the pores at day 4 in culture. After 14 days in culture, the cell clusters appear as one particle. The hepatocytes maintain their biological function when grown as aggregates within the alginate sponges. Albumin secretion rates from hepatocytes seeded within the alginate sponges is considerably higher than those obtained for cells grown on standard collagen type I and continues for longer times. Urea synthesis by the hepatocytes immobilized in the alginate sponges remains relatively high at the end of 15 days. Studies are now in progress to evaluate these new cell constructs, in vivo.

Keywords: alginate sponges, cell culture, cell scaffolds, hepatocytes.

Introduction

Liver transplantation is currently the main mode of treatment for patients with acute fulminant hepatic failure [1]. The main problems that arise from this situation are donor shortage, and logistics of matching the appropriate donor with the patient in need at the right time. Thus, only 20% of the candidates reach orthotopic liver transplant (OLT), and the others are at risk of death. Temporary liver support may save the life of these people by bridging until a donor liver is found [2]. In certain situations of acute liver failure, the temporary hepatic sup-

---

Address for correspondence: Prof Jose C. Merchuk, Head, Unit of Biotechnology, Ben-Gurion University of the Negev, P.O. Box 653, Beer-Sheva 84105, Israel. Tel.: +972-7-6461-768. Fax: +972-7-6472-916. E-mail: jcm@bgumail.bgu.ac.il

port is very important per se, since it can give the time needed for liver regeneration without recurring to transplantation.

The approaches used presently for hepatic support are based on hemodialysis, hemoperfusion, hemofiltration and plasmaphoresis. Recently, the FDA approved an extracorporeal device for ex vivo blood purification, using an animal liver or hepatocytes cultured within an artificial liver device. An additional approach that is rigorously investigated is liver cell transplantation [3]. The source of the cell transplant may be genetic engineered cells, or hepatocytes obtained by dissociation of the donated tissue. The cells may be injected as a suspension, or transplanted after being processed and reorganized as a bioartificial tissue by the tissue engineering approach.

The approach of tissue engineering relies on a three-dimensional scaffold to guide the dissociated cells into forming a new functioning tissue. The procedure for tissue engineering of liver tissue involves: 1) designing and manufacturing an optimal polymeric scaffold; 2) taking a small piece of liver from a donor and isolating the hepatocytes; 3) seeding the hepatocytes within the polymeric scaffold to guide their organization into a functioning unit; and 4) implanting the scaffold containing the hepatocytes in the patient. A variation being considered on this last step is the prior implantation of the scaffold in patients for vascularization, and latter injection of the cells in the prevascularized matrix. In the present paper, we review our efforts to develop thick scaffolds with a structure to match the needs of hepatocytes.

The alginate scaffold

The polymeric matrix serving as a scaffold for artificial liver tissue should have the following properties: 1) be biocompatible, in order to avoid tissue reaction after implantation; 2) have an appropriate pore size, in order to allow an efficient cell penetration, and the vascularization in the patient's body after implantation; 3) have high porosity to allow the seeding of a large cell mass; 4) have interconnected pores, to facilitate cell reorganization into a tissue-like functioning unit; 5) have appropriate mechanical properties to assure flexibility and structure conservation, and to avoid collapse upon implantation; and 6) be bioresorbable, so that it will disappear when no longer in need.

Recently, our group has produced a new alginate scaffold [4] that appears to have the above-mentioned properties, and in addition is hydrophilic and therefore easily wetted by the culture medium. Thus, cell seeding into the matrix pores should be more efficient, leading to a better distribution of the cells in the implant volume. This new scaffold may be considered both as a matrix for cell transplantation, and as a support for three-dimensional culture to be used in extracorporeal devices.

"Alginate" refers to a family of polyanionic copolymers derived from brown sea algae and comprising 1,4-linked β-D-mannuronic (M) and α-L-guluronic acid (G) residues in varying proportions. It is soluble in aqueous solutions, at room temperature, and forms stable gels in the presence of certain divalent cations, such as calcium, barium, and strontium. The unique properties of alginate, combined with its biocompatibility [5,6], hydrophilicity, and relatively low cost have made alginate an important polymer in pharmaceutical applications (e.g., wound dressings, dental impression material). Alginates of a pharmaceutical grade and which comply with all the quality requirements in the European and US Pharmacopoeias can be obtained from several manufacturers. The use of alginate material for cell encapsulation has been limited to hydrogels or as part of the polyelectrolyte complex in the immunoisolation membrane. In these applications, the porosity of the gels is in the range of 10 μm and below, and this does not allow enough space for growth and three-dimensional rearrangement. The new method [4] makes possible to produce alginate sponges with much larger pores. The shape and size of these pores can be set by the fabrication method [7].

Sponge preparation
Alginate sponges were prepared by a gelation-freeze dry technique [4] which consists of the following steps: 1) preparation of 2% (w/v) stock solution of sodium alginate; 2) cross-linking the alginate in the presence of the bivalent cation, calcium ions; 3) freezing the gels; and 4) lyophilization to obtain the sponges.

Sponge morphology by scanning electron microscopy (SEM)
The alginate sponges displayed a highly porous morphology with pore structure (pore size and wall thickness) dependent on the type and concentration of alginate and cross-linker, and the freezing regime [4,7]. In particular the freezing regime, i.e., liquid nitrogen ($-180°C$), freezer ($-20°C$), or oil bath ($-35°C$), affected the pore shape and size distribution throughout the sponge [7]. Figure 1 shows a typical scanning electron micrograph of a horizontal cut at the top of a sponge fabricated using the liquid nitrogen freezing technique. The liquid nitrogen-type sponges displayed a gradient in pore size, with small, closely packed pores on the bottom and large, elongated pores at the top. The average pore size at the top and bottom part of the sponge was 150 and 55 μm, respectively. The freezer-type sponge, on the other hand, displayed a larger, more uniform pore size (130-μm pore size) with no preferential orientation of the pores (Fig. 2). Sponges made by the oil bath freezing technique showed similar morphology to those made by the liquid nitrogen technique.

To investigate the cause of these differences, the temperature at various points of the alginate gel, while immersed in an oil bath, was measured using two temperature probes [7]. The results indicate an initial temperature difference of about $4°C$ between the top and bottom of the gel. This is explained by the fact that the cooling medium interfaces the plate primarily at its bottom surface, inducing a temperature gradient within the gel. This cooling method produced sponges with small, closely packed pores at the bottom of the sponge and an upper layer of wider, elongated pores, poised in a vertical manner. This morphology can be explained by the different rate of nucleation of the ice crystals in the two opposite

122

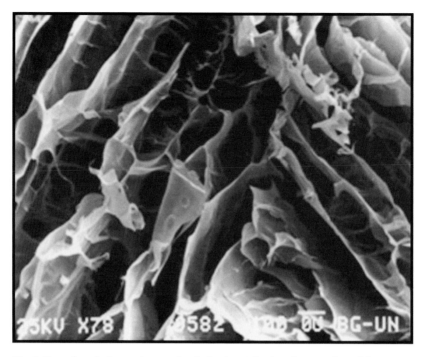

Fig. 1. Scanning electron micrograph of a horizontal cut at the top of an alginate sponge fabricated using the liquid nitrogen technique. The alginate used was Protanal LF 120 (Protan Inc.) and calcium gluconate as the cross-linker (From [7]).

sides of the gel. The bottom layer is rapidly cooled to subfreezing temperatures so that only very small ice crystals form. During the crystallization process, the cross-linked alginate molecules are forced to the perimeter of the crystal, thus forming the pore structure. The upper part of the gel is cooled at a slower pace, allowing larger crystals to be formed, which grow from the cooler bottom to the top. Though no temperature measurements were conducted during the liquid nitrogen and freezer procedures (for technical reasons), a temperature profile in the gels during processing can be presumed. The liquid nitrogen cools the plate from its bottom, producing a gradient in pore size. The freezer, on the other hand, provides a more homogenous cooling environment, and the cooling rate is slower. Thus, the crystals formed are larger and homogeneously distributed throughout the sponge. This should lead to sponges with larger pores and with a more uniform pore size throughout the matrix, as indeed shown in Fig. 2.

Sponge stability in medium
Alginate sponges, wetted with the culture medium, maintained their porous morphology for over a month [7]. Their overall appearance differed from that of alginate hydrogel of the same composition, which did not undergo lyophilization. Moreover, the sponges retained their shape and volume over prolonged times in

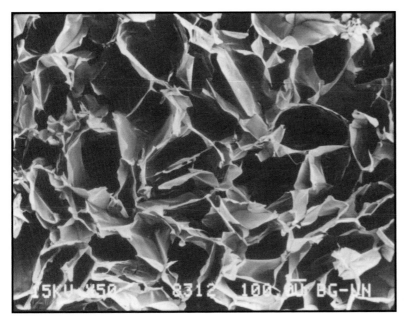

Fig. 2. Scanning electron micrograph of a horizontal cut at the top of a sponge fabricated using the freezer technique. The alginate and cross-linkers used are the same as indicated in Fig. 1 (From [7]).

medium, at 37°C, while the unlyophilized cross-linked alginate dissolved relatively fast, within 1 week in the culture medium. The enhanced stability of the sponges is attributed to the combined effect of cross-linking and the lyophilization process (without cross-linking, the sponge loses its shape and volume upon wetting with growth medium). It is not yet clear whether the stabilizing effect of the lyophilization process is chemical or physical in nature. Studies are underway to investigate this point.

Hepatocytes culture within alginate sponges

Our initial studies focused on evaluating the newly designed alginate sponges as hepatocyte scaffolds [8,9]. Four criteria were used to assess hepatocyte-sponge interactions: cell distribution within the sponge; cell longevity within the sponge; morphology of the cells; and retention of hepatospecific functions of the seeded cells. The hepatocytes were isolated from male Sprague Dawley rats and were cultured in a chemically defined serum-free medium (William's E with 10 ng/ml EGF, 20 mU/ml insulin, 5 nM dexamethasone, 20 mM pyruvate, 2 mM L-glutamine, 100 U/ml penicillin/streptomycin, and 0.2 mM gentamycin). The cells were seeded within sponges, at a cell density of 1×10^6 cells/sponge, and incubated in a 5% CO_2 incubator, at 37°C, with 99% humidity. The alginate sponges used in the tissue culture studies were made of alginate Protanal LF 120 (Protan Inc.) and calcium gluconate as the cross-linker, and were frozen by the freezer technique.

They were prepared in a 24-well plate with the following dimensions: 15×10 mm (d × h) and volume of 1.77 cm^3 and were sterilized using ethylene oxide gas.

Hepatocyte longevity within alginate sponges

Hepatocytes were seeded at a density of 1×10^6 cell per sponge. After 1 day in culture, triplicate sponges were sacrificed and the number of cells was analyzed by the MTT assay, which measures the ability of mitochondrial dehydrogenase enzymes to convert the soluble yellow MTT salt (3(4,5-dimethyl-thiazol-2-yl)-2,5-tetrazolium bromide) into insoluble purple formazan salt. The 1-day samples yielded $0.9-1 \times 10^6$ cells/sponge, indicating that all seeded cells are efficiently entrapped within the sponges [8,9]. One day after seeding, the cells are seen homogeneously dispersed all over the sponge volume and are located mostly within the pores.

Overall, the hepatocytes in sponges maintained their viability throughout the culture period (3 weeks−1 month) as detected by the viable-dead staining method, where the cells are dual-stained with two fluorescent dyes, DiOC$_{18}$ and propidium iodide. The first dye stains the cell membrane in green, and the second dye stains the nucleus of dead cells in red. Visual examination of the sponges at different focal points revealed that most of the cells are green, even after 1 month in culture [9]. The prolonged longevity of the cells is further established when comparing the cell number as determined by the MTT assay to that assessed by the total DNA (which gives the total cell count, independent of their viability). The good correlation between the two assays, as shown in Fig. 3, indicates that the cells in sponges maintain their viability for at least 12 days in culture. Moreover, the practically constant cell count as measured by the DNA method indicate that cell loss during the daily medium change is negligible. After 12 days in culture, the MTT results show a decrease in cell viability. This may relate to cell death or to a decrease in enzyme activity due to changes in cell morphology and behavior during the prolonged culture. At present, we are conducting studies to verify this point. The retention of cell viability within alginate sponges as well as the capability of these sponges to physically retain the cells indicate their suitability as scaffolds for hepatocyte culture.

Hepatocyte morphology within alginate sponges

The morphology of hepatocytes within the sponges as a function of culture time was followed using SEM. Typical micrographs can be seen in Fig. 4 [8]. Within 2 days after seeding, the cells begin to aggregate within the pores indicating that the cells prefer interacting with their neighboring cells than with the matrix. The aggregate size increased with time until reaching a stationary size at day 4. The rearrangement of the dissociated hepatocytes into aggregates indicates that the pores are interconnected, enabling the cells to move from one pore to the other. After 14 days in culture, the cell clusters appear as one particle, in which the cells cannot be distinguished from each other.

Figure 5 shows the change in the mean diameter of the hepatocyte aggregates

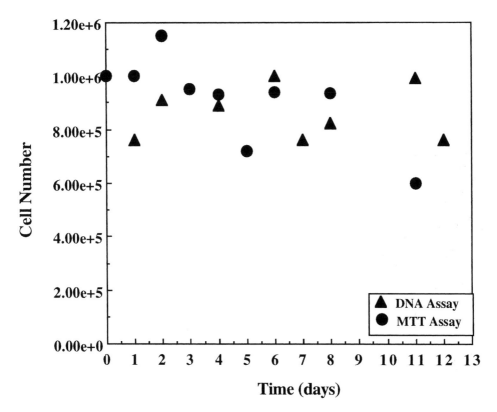

Fig. 3. Cell number in alginate sponges estimated by the DNA test for total cell count and MTT assay for viable cells, over a period of 12 days. Type of sponges as indicated in Fig. 2 (From [9]).

with time, for cells cultured under different conditions [9]. When grown in sponges, hepatocyte aggregates reach their maximal average size, which is approximately 80 µm, after 4 days in culture. This value is closely related to the average pore size of sponges made by the freezer technique (~ 100 µm). It is apparent that the size of the spheroids is controlled by the mean pore diameter of the sponge. This is further supported if the results are compared to those obtained in T-flasks and in a spinner flask [9]. The size of the spheroids generated in the spinner is twice that in the sponges, and the T-flask produces aggregates 8 times larger than that in the sponges. It seems that in the case of the spinner, the spheroid size is mainly controlled by the shear in the system.

Figure 6 shows the size distribution of the hepatocyte spheroids at different sponge depths [9]. The measurements were made on cross-sections of a sponge with a slice thickness of 200 µm. The figure shows that the aggregates are evenly distributed in the entire volume of the sponges, almost up to the bottom of the sponges. The average diameter of the spheroids throughout the sponge is approximately 100 µm. This is in agreement with the uniform pore size structure of the sponge used in these studies (~ 150 µm).

126

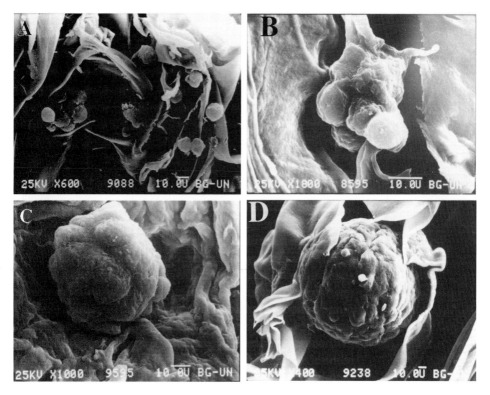

Fig. 4. Scanning electron micrograph showing the kinetics of hepatocyte aggregation within the alginate sponge. **A:** Immediately after cell seeding, and after 4 (**B**), 11 (**C**), and 14 (**D**) days in culture. Type of sponges as indicated in Fig. 2 (From [9]).

The aggregation of the hepatocytes into spheroid clumps has been previously observed when grown in suspension under mixing or when plated on positively charged petri dishes. This phenomenon is well-documented both in morphology and ultrastructure [10,11]. Spheroids have also been obtained when growing hepatocytes in porous supports [12–14]. Several schemes for bioartificial liver have been proposed based on such spheroids, claiming that this self-arrangement enhances liver-specific functions [12,13]. Yanagi et al. [13] obtained a very high density of hepatocytes (1.2×10^7 cells/ml) immobilized in collagen-coated, reticulated polyvinyl formal (PVF) foam, and reported also the presence of such spheroids. On the other hand the report by Miyoshi et al. [14] shows long-term performance of albumin secretion of hepatocytes cultured in a packed bed reactor utilizing PVF, while most of the hepatocytes remained as single cells, mainly round-shaped. They concluded that spheroid formation is not necessarily required if the cells are packed densely enough for the construction of a three-dimensional structure of individual cells. In all cases, it is recognized that cell organization in some level is required in order to maintain the hepato-specific functions.

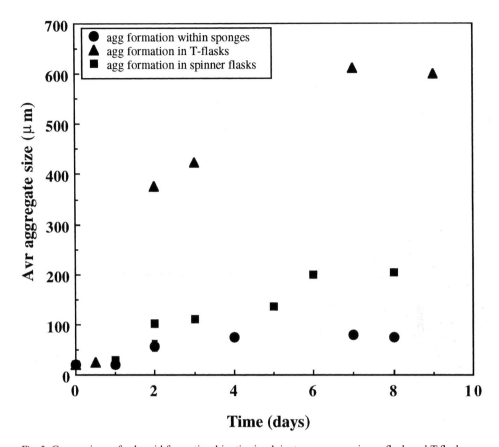

Fig. 5. Comparison of spheroid formation kinetics in alginate sponges, spinner flask and T-flasks.

Hepatocyte function within alginate sponges

A series of experiments was run in order to study to what extent the hepatocytes maintained their biological function when grown as aggregates within the alginate sponges [8,9]. Experiments were run in parallel with hepatocytes cultured as monolayers on collagen type I. The results presented in Fig. 7 reveal that the albumin secretion rates from hepatocytes seeded within the alginate sponges were considerably higher than those obtained for cells grown on standard collagen type I films. Moreover, the albumin secretion process continued for longer. While albumin secretion decreased nearly to zero within 5 days in cells seeded on collagen type I, cells in sponges maintained high secretion for over 2 weeks. The high-secretion rate from hepatocytes in the alginate sponges is related to their spheroid morphology, i.e., their organization into a three-dimensional tissue-like structure. In contrast, the hepatocytes grown on the collagen-coated plates showed a flat morphology.

Urea synthesis by the hepatocytes immobilized in the alginate sponges is shown in Fig. 8. It can be seen that the activity remains relatively high at the end of 15

128

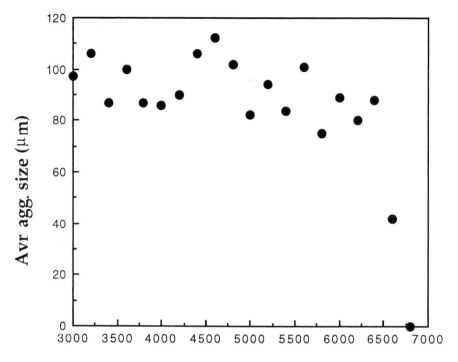

Fig. 6. Distribution of spheroids in the volume of an alginate sponge. The abscise is the distance of the upper part of the sponge (From [9]).

days, and is comparable to data on urea synthesis by hepatocytes immobilized within collagen-coated PVF [14].

Recapitulation

In this paper, a novel type of scaffold for hepatocyte culture and transplantation has been presented. The scaffold consists of a sponge fabricated from alginates by a multistep technique that includes gelation, freezing and lyophilization. The scaffold is characterized by a highly porous structure with interconnecting pores, making it suitable for tissue engineering purposes and as a support for three-dimensional culture to be used in extracorporeal devices. The hydrophilic nature of the sponge results in an easy wetting by the culture medium and an efficient cell seeding in the internal volume of the scaffold.

Experiments run with the alginate sponges showed that a very good distribution of cells in almost the entire volume of the scaffold was obtained, when using a scaffold of approximately 1.77 cm^3 (15 mm diameter and 10 mm high). This is a much larger volume for a scaffold than those used in other published papers [14,15], which typically have a thickness smaller by one order of magnitude. The

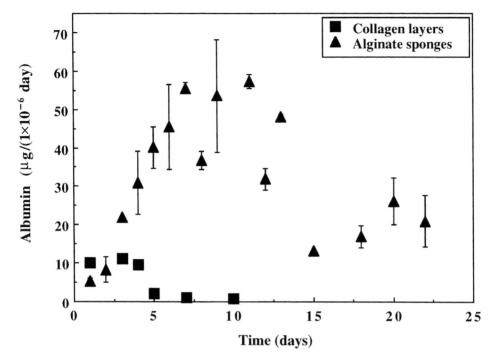

Fig. 7. Comparison of albumin secretion rates by hepatocytes grown as monolayers on collagen or within alginate sponges. Type of sponges as indicated in Fig. 2 (From [9]).

ability to handle a thick specimen is important especially when considered for liver cell transplantation, where a large cell mass is needed to replace the multiple functions of a damaged liver. Thick specimens are also advantageous as they allow easier handling in the surgical procedure. On the other hand, mass transfer problems are expected in such a system unless a very rapid vascularization inside the scaffold is attained.

It is widely accepted that the best results in terms of cellular function (i.e., albumin secretion, urea synthesis) are achieved when the hepatocytes in culture are aggregated as spheroids. The use of the present support does not require previous preparation of spheroids in Petri dishes before seeding in the scaffold [11]. The hepatocytes are seeded as dispersed cells into the sponge and within a short time form the spheroids. It appears that the hepatocytes, even dispersed as single cells, preserve the necessary information to reconstruct a proper three-dimensional cytoarchitecture that creates the appropriate microenvironment required for the basic hepatic functions [15]. The hydrophilic properties of the alginate sponges and their unique morphology allow the homogeneous distribution of the cells, and their rearrangement into aggregates. The uniform porous structure of the sponges also allows the distribution of the hepatocytes in the entire volume of the scaffold. This distribution remains homogeneous even after the formation of the spheroids.

130

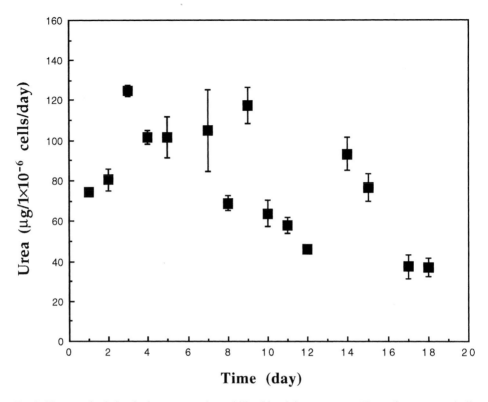

Fig. 8. Urea synthesis by the hepatocytes immobilized in alginate sponges. Type of sponges as indicated in Fig. 2. (From [9].)

The prolonged viability of hepatocytes in the scaffold and the maintenance of their hepatic functions indicate that the new alginate scaffold is capable of offering an appropriate support for prolonged liver-specific functions. The challenge remaining ahead is the increase in the loading of hepatocytes in the sponge, and the attainment of rapid vascularization upon implantation. An alternative approach is the implantation of the sponge for prevascularization, and *a posteriori* seeding of the hepatocytes, once a sufficient mass exchange rate is ensured. Work in this direction is in progress.

Acknowledgements

The authors acknowledge the support of The Israeli Ministry of Science and The Harry Stern Applicative Fund.

References

1. Langer R, Vacanti JP. Tissue engineering. Science 1993;260:920–926.
2. Takahashi T, Malchesky PS, Nose Y. Artificial liver: state of the art. Dig Dis Sci 1991;36:

1327—1333.

3. Vacanti JP, Morse MA, Zaltsman WM, Domb AJ, Perez-Altayde A, Langer R. Selective cell transplantation using bioresorbable artificial polymers as matrices. J Pediatr Surg 1988;23:3—9.

4. Shapiro L, Cohen S. Novel alginate sponges for cell culture and transplantation. Biomaterials 1997;18:583—590.

5. Sennerby L, Rostlund T, Albrektsson B, Albrektsson T. Acute tissue reactions to potassium alginate with and without colour/flavor additives. Biomaterials 1987;8:49—52.

6. Cohen S, Bernstein H, Hewes C, Chow M, Langer R. The pharmacokinetics of and humoral responses to antigen delivered by microencapsulated liposomes. Proc Natl Acad Sci USA 1991; 88:1040—1044.

7. Zmora S, Glicklis R, Cohen S. Cell scaffolds based on alginate sponges. In: Peppas NA, Mooney DJ, Mikos AG, Brannon-Peppas L (eds) Proceedings of the Topical Conference on Biomaterials, Carriers for Drug Delivery, and Scaffolds for Tissue Engineering, AIChE. Los Angeles, 1997;84—86.

8. Glicklis R, Zmora S, Agbaria R, Cohen S. The interaction of hepatocytes with 3-D porous alginate scaffolds. Proc. 2nd World Meeting APGI/APV, Paris, 1998.

9. Glicklis R, Zmora S, Agbaria R, Cohen S. Hepatocyte culture in 3-D sponges, submitted.

10. Peshwa MV, Wu FJ, Follstad BD, Cerra FB, Hu W-S. Kinetics of formation of hepatocyte spheroids. Biotech Prog 1994;10:460—466.

11. Wu FJ, Peshwa MV, Cerra FB, Hu W-S. Entrapment of hepatocyte spheroids in a hollow fiber bioreactor as a potential bioartificial liver. Tissue Eng 1995;1:29—40.

12. Matsushita T, Ijima H, Koide N, Funatsu K. High albumin production by multicellular spheroids of adult rat hepatocytes formed in the pores of polyurethane foam. Appl Microbiol Biotechnol 1991;36:424—426.

13. Yanagi K, Miyoshi H, Fukuda H, Ohshima N. A packed-bed reactor utilizing porous resin enables high density culture of hepatocytes. Appl Microbiol Biotechnol 1992;37:316—320.

14. Miyoshi H, Yanagi K, Fukuda H, Ohshima N. Long term performance of albumin secretion of hepatocytes cultured in a packed bed reactor utilizing porous resin. Artif Organs 1996;20: 803—807.

15. Kasai S, Sawa M, Mito M. Is the biological artificial liver clinically applicable? A historic review of artificial liver support systems. Artif Organs 1994;18:248—354.

Engineering of cell lines for diabetes therapy

Christopher B. Newgard[1], Christian Quaade[2], Anice Thigpen[2], Hans E. Hohmeier[1,2], Veronique Vien Tran[1], Fred Kruse[2], He-Ping Han[2], George Schuppin and Samuel Clark[2]

[1]Gifford Laboratories for Diabetes Research and Departments of Biochemistry and Internal Medicine, University of Texas Southwestern Medical Center at Dallas, Dallas; and [2]BetaGene, Inc., Dallas, Texas, USA

Abstract. Unique challenges must be confronted by tissue engineers seeking to develop cell-based insulin replacement strategies for treatment of type I diabetes. Isolation and transplantation of islets of Langerhans from human or large animal sources has been the traditional approach, but it is questionable whether these methods will provide a reliable source of tissue for large numbers of patients with the disease. A potential alternative to transplantation of primary cells may be development of cell lines that can be grown in large quantities at relatively low cost, but only if such cells can secrete insulin in response to appropriate physiological cues. This article summarizes progress made by our group in developing rodent cell lines that mimic the function of the normal islet β-cell and that are protected from specific components of the immune response. Practical application of these technologies may ultimately require development and engineering of human cell lines, using the lessons learned from animal experiments summarized in this article.

Keywords: cell lines, genetic engineering, insulin, molecular biology, transplantation.

Introduction

As summarized in this compendium of articles, tissue engineering refers to the application of molecular and/or cell biological tools to create a replacement for tissue that is destroyed or rendered nonfunctional by human disease or injury. This involves development of cells that can perform and remain viable in the context of biocompatible synthetic materials that provide structural integrity, partial immunoprotection, and a suitable matrix for cell adherence and growth. In the case of a cell-based therapy for diabetes, which is the focus of this particular chapter, these engineering goals also apply. However, an added complexity for the construction of an "artificial pancreas" is the need to replicate faithfully the capacity of normal insulin producing β-cells of the islets of Langerhans to secrete the hormone in a precisely regulated fashion. The islet β-cell is unique in its capacity to secrete its hormonal product, insulin, in direct proportion to the concentration of glucose in the blood. This occurs because the rate of glucose

Address for correspondence: Christopher B. Newgard PhD, Gifford Laboratories for Diabetes Research, University of Texas Southwestern Medical Center, 5323 Harry Hines Blvd., Dallas, TX 75235, USA. Tel.: +1-214-648-2930. Fax: +1-214-648-9191.

metabolism is increased in the β-cell as blood glucose is raised, the release of insulin occurring by virtue of generation of metabolic "stimulus/secretion coupling factors" (reviewed in [1]). While the metabolic signals that link glucose metabolism to insulin secretion are incompletely understood, one important aspect of the response appears to be an increase in the ATP:ADP ratio, resulting in closure of ATP sensitive K^+ channels and consequent opening of voltage-gated Ca^{2+} channels. What is also clear is that the rate of glucose metabolism in islet β-cells is regulated to a large measure by the activity of the glucose phosphorylating enzyme glucokinase [2,3]. Rapid equilibration of extracellular and intracellular glucose concentrations in β-cells is assured by expression of the high-capacity GLUT-2 glucose transporter. GLUT-2 and glucokinase have been described as central components of the "glucose sensing apparatus" of islets, and are important tools for enhancing the function of candidate therapeutic insulin secreting cell lines [3].

Transplantation of isolated islets of Langerhans has been investigated as an alternative to whole organ transplantation for more than three decades, yet only a small number of patients have been successfully treated by this approach. Major issues that remain to be overcome in order for cell-based insulin replacement in insulin dependent diabetes mellitus (IDDM) to become widely applicable include: 1) development of a replenishable, inexpensive source of cells; 2) development of a source of cells that maintain a stable phenotype both in vitro and in vivo; and 3) immunoprotection of transplanted cells. Because of the difficulty and cost associated with islet isolation and pancreas transplantation, there has been increasing emphasis on application of the tools of molecular biology for engineering of insulin secreting cell lines with fuel-mediated insulin secretion responses resembling those of normal β-cells, with the hope that such lines might serve as a surrogate for islets in transplantation therapy. This article summarizes work from our group in this area.

Towards a solution to the problem of cell supply

Islet harvesting requires digestion of the pancreas with collagenase, and separation of islets from other cells and tissue by mechanical devices or gradient centrifugation. This procedure yields between 200,000 and 1 million islets per human pancreas, while the number of islets required for treatment of diabetes ranges from 0.5 to 2.0×10^6 islet equivalents per patient [4]. Thus, transplantation of human islets does not represent a broadly applicable therapy for the millions of individuals with IDDM worldwide. Harvesting of porcine islets has been suggested as an alternative strategy, but significant obstacles remain such as development of strategies for protecting porcine islet xenografts from immune destruction and methods for preparing large lots of functionally equivalent cells.

In light of these issues, we and others have been investigating the utility of immortalized cell lines as vehicles for insulin replacement in IDDM. The idea is that appropriately modified cell lines could solve the issue of cell supply, if they can be grown in large quantity without loss of key functional properties. To date, no sustainable human insulinoma cell lines have been developed, although several groups, including our own, are actively engaged in attempting to procure them by a variety of methods. In the interim, relevant engineering strategies have been developed using rodent cell lines.

The approach being developed by our group involves use of the tools of molecular biology to enhance the function of pre-existing cell lines. The goal has been to increase insulin content in candidate cell lines by overexpression of the human proinsulin cDNA, and to ensure that insulin secretion is appropriately regulated by glucose and other potentiators of insulin secretion. Early experiments of this type were carried out in the neuroendocrine cell line AtT-20ins, which is derived from ACTH-secreting cells of the anterior pituitary. When transfected with the human proinsulin cDNA, these cells synthesize, process, and secrete the mature insulin polypeptide [5,6]. While insulin secretion from AtT-20ins cells is not normally responsive to glucose [7], stable transfection with a plasmid containing the GLUT-2 glucose transporter conferred this property [8]. However, transplantation of engineered AtT-20ins cells into nude rats caused insulin resistance, apparently due to ACTH secretion from the cells and resultant adrenal hyperplasia [9].

Our more recent work has instead focused on the rodent insulinoma cell line RIN1046-38, which is able to respond to glucose at low passage numbers, but loses both glucose sensing and expression of GLUT-2 and glucokinase with time in culture [10,11]. Among available insulinoma cell lines, RIN 1046-38 cells are intermediate in terms of their phenotypic resemblance to normal islets, being much more differentiated than RINm5F cells, but exhibiting lower insulin content and less glucose responsiveness than other lines such as MIN-6, INS-1, or βTC-6 [12–14]. There are two major reasons for choosing this cell line as our template for engineering studies. First, RIN1046-38 cells were derived from a radiation-induced tumor rather than by overexpression of an oncogene such as SV-40 T-antigen, and as such have remained stably dipoloid for more than 20 years; evidence of genetic stability. Consistent with this, glucagon and somatostatin expression is essentially undetectable in these cells, in contrast to other cell lines in which these genes have been reported to increase in expression with time in culture. Second, based on experiences to date, it is likely that any human β-cell lines that can be developed will be substantially dedifferentiated compared to primary islet cells [15]. Thus, it is hoped that development of workable strategies for engineering of moderately differentiated rodent insulinoma cell lines will be rapidly applicable to improving the function of candidate human cell lines.

Using a process of "iterative engineering", or the step-wise stable transfection of cells with plasmids containing discrete antibiotic resistance genes, we have recently developed rodent insulinoma cell lines with insulin content comparable to that measured in cultured human islet preparations [16]. We have also stably installed the GLUT-2 and glucokinase genes in these cells, and shown that this maneuver confers a large enhancement in glucose-stimulated insulin secretion

[16,17]. The multiply engineered lines increase insulin secretion by 6- to 8-fold in response to glucose, effects that are further potentiated in a fashion similar to that seen in β-cells by agents that raise cAMP [17]. These responses are clearly larger than observed in unengineered cells, and perfusion studies reveal that insulin secretion is sustained in glucokinase and/or GLUT-2 expressing cells more effectively than in unengineered cells or cells that express the human insulin transgene alone [17]. The engineered cell lines exhibit a maximal response to glucose at low concentrations (0.05—0.25 mM), but can be shifted such that their maximal response occurs at 3—5 mM glucose by performing secretion studies in the presence of 2 mM 5-thioglucose, which inhibits low Km hexokinase activity and glucose usage by approximately 30% [17]. These data suggest that further adjustment of glucose sensing to simulate the response threshold of the β-cell (4—5 mM) should be achievable by stable suppression of low Km hexokinase activity. We are currently investigating several strategies for achieving this goal, including gene "knock-out" by homologous recombination.

One advantage of the stable transfection approach outlined above is that it may provide a remarkable genetic and phenotypic stability in the engineered cell lines. Consistent with this, transplantation of unencapsulated engineered RIN cell lines into nude rats reveals that stably integrated GLUT-2, glucokinase, and human insulin transgenes under control of the cytomegalovirus (CMV) or rat insulin (RIP) promoters are expressed at constant levels in the in vivo environment over the full duration of experiments performed to date (48 days) [16]. Several endogenous genes expressed in normal β-cells, including rat insulin, amylin, sulfonylurea receptor, and glucokinase are stably expressed in the insulinoma lines during these in vivo studies. Further, insulin secretion in response to glucose and other secretagogues is stably maintained in vitro in periodically monitored, continuously cultured cells for at least 4 months (the longest period tested to date) (unpublished observations). These results suggest that a highly important criterion for choosing human β-cell lines that might serve as a vehicle for insulin replacement will be genetic stability. If such a cell line can be identified as the starting template, the engineering approaches outlined in this article and described in more detail in the referenced studies should allow the procurement of therapeutically useful lines.

Immmunoprotection

As new rodent cell lines with attractive physiological properties are being developed, insights into strategies for immunoprotecting such cell lines are also emerging. As is the case with strategies for engineering of physiological responsiveness, work to date on immunoprotection has been limited to animal cell lines, as appropriate human β-cell lines are not yet available. However, it is hoped that emerging approaches will be rapidly transferable to human cells as they are developed.

Immunological destruction of β-cells in IDDM is thought to be a T-cell-dependent process. Infiltration of pancreatic islets by mononuclear cells of the immune system, mostly macrophages and T lymphocytes, precedes β-cell destruction in human subjects with IDDM and in nonobese diabetic (NOD) mice. Impaired function and destruction of β-cells appears to result from direct contact with islet-infiltrating cells and/or exposure to inflammatory cytokines that they produce [18—21]. Since it is thought that cytotoxic T-cells destroy target cells via docking to MHC class I molecules, resulting in T-cell receptor activation, one approach for providing partial protection against immunological damage of transplanted islets or engineered cell lines is encapsulation of such cells in perm-selective materials such as alginate or synthetic membranes. These maneuvers have been demonstrated to slow destruction of islet xenografts or cell lines transplanted as allografts. Another potential advantage of implantation of cells in an encapsulated form, particularly when all the cells are within a single device (macroencapsulation), is the immediate localization or retrieval of cells should problems with cell performance or other complications be encountered. However, it is unlikely that permselective membranes or devices will be sufficient to provide full protection for insulin producing cells when transplanted into individuals with fulminant autoimmune diabetes. Such devices must be designed to allow efficient export of insulin, conditions which will unfortunately also allow entry of cytokines and chemical toxins such as nitric oxide or oxygen radicals.

Based on these assumptions, our approach to engineering of immunoprotection in insulin-producing cell lines has focused on blocking the cytotoxic effects of small molecular weight mediators of the immune response such as cytokines and toxic oxygen and nitrogen radicals. We found that the cytokine interleukin-1β (IL-1β) causes destruction of INS-1 insulinoma cells, while having no effect on RIN 1046-38 and its engineered derivatives, and that this difference is correlated with a higher level of expression of manganese superoxide dismutase (MnSOD) in the latter cells [22]. Stable overexpression of MnSOD in INS-1 cells provides complete protection against IL-1β mediated cytotoxicity, and also results in markedly reduced killing when such cells are exposed to conditioned media from activated human or rat peripheral blood mononuclear cells (PBMC). Further, overexpression of MnSOD in either RIN or INS-1-derived lines results in a sharp reduction in IL-1β-induced nitric oxide (NO) production, a finding that correlates with reduced levels of the inducible form of nitric oxide synthase (iNOS). Treatment of INS-1 cells with L-NMMA, an inhibitor of iNOS, provides the same degree of protection against IL-1β-mediated cell killing as MnSOD overexpression, supporting the idea that MnSOD protects INS-1 cells by interfering with the normal IL-1β-mediated increase in iNOS [22]. Since NO and its derivatives have been implicated as critical mediators of β-cell destruction in IDDM, insulinoma cell lines engineered for MnSOD overexpression may be less susceptible to the immune response when transplanted into such patients. However, we also noted that MnSOD overexpression failed to provide protection in either RIN or INS-1 cells against the cytotoxic effects of γ-interferon (γ-IFN). We are currently investigating the apparently divergent pathways of IL-1β and γ-IFN-mediated destruction of insulinoma cells and are developing strategies for blockade of the γ-IFN pathway.

Conclusions and Future issues

Important gains have been made in recent years in identifying and developing cell lines that can perform key functions of the normal islet β-cell, and in devising strategies that may aid in protecting such cells when transplanted into an immunocompetent host. It is hoped that experience with the cellular engineering approaches applied thus far to rodent cell lines will be applicable in the near future to development of cell lines suitable for treatment of human diabetes. It is as yet uncertain if rodent cell lines will serve as a useful vehicle for delivery of insulin to human patients. Problems with such cell lines include their natural production of rodent insulin, and the possibility that release of this and other xenoantigens can lead to destruction of the graft, even in cells engineered with immunoprotective genes. It is therefore imperative to develop human cell lines that can serve as functional allografts. This task is made difficult by the relative resistance of human cells to transformation, and by the limitations in experimental maneuvers that can be applied to our species. Nevertheless, efforts at producing such cell lines are a high priority, since it is likely that the marriage of these cells with the emerging engineering technologies will be required in order to make cell-based insulin replacement a practical reality for large numbers of patients with IDDM.

Acknowledgements

Studies relevant to this article performed at the Gifford Laboratories for Diabetes Research at the University of Texas Southwestern Medical Center at Dallas were supported by grant R01-DK-46492 from the National Institutes of Health, by a National Institutes of Health/Juvenile Diabetes Foundation International Diabetes Interdisciplinary Research Program, and by a grant from the Redfern Foundation, Midland, Texas.

References

1. Newgard CB, McGarry JD. Stimulus/secretion coupling in the islet β-cell. Ann Rev Biochem 1995;44:266–301.
2. Meglasson M, Matschinsky FM. The glucose sensor hypothesis. Diabet Metab Rev 1986;20: 696–745.
3. Newgard CB. The role of glucose transport and phosphorylation in control of insulin secretion. Diabet Rev 1996;18:235–277.
4. Sutherland DER, Gores PF, Hering BJ, Wahoff D, McKeehen DA, Guressner RWG. Islet transplantation: an update. Diabet/Metab Rev 1996;12:137–150.
5. Moore H-P, Walker MD, Lee F, Kelly RB. Expressing a human proinsulin cDNA in a mouse ACTH-secreting cell. Intracellular storage, proteolytic processing, and secretion on stimulation.

Cell 1983;35:531—538.

6. Gross DJ, Halban PA, Kahn RC, Weir GC, Villa-Komaroff L. Partial diversion of a mutant proinsulin (B10 aspartic acid) from the regulated to the constitutive secretory pathway in transfected AtT-20 cells. Proc Natl Acad Sci USA 1989;86:4107—4111.

7. Hughes SD, Quaade C, Milburn JL, Cassidy LE, Newgard CB. Expression of normal and novel glucokinase mRNAs in anterior pituitary and islet cells. J Biol Chem 1991;266:4521—4530.

8. Hughes SD, Johnson JH, Quaade C, Newgard CB. Engineering of glucose-stimulated insulin secretion and biosynthesis in non-islet cells. Proc Natl Acad Sci USA 1992;89:688—692.

9. BeltrandelRio H, Schnedl WJ, Ferber S, Newgard CB. Genetic engineering of insulin secreting cell lines. In: Lanza RP, Chick WL (eds) Pancreatic Islet Transplantation, Volume 1: Procurement of Pancreatic Islets. Austin, TX: R.G. Landes Co., 1994;169—183.

10. Clark SA, Burnham BL, Chick WL. Modulation of glucose-induced insulin secretion from a rat clonal β-cell line. Endocrinology 1990;127:2779—2788.

11. Ferber S, BeltrandelRio H, Johnson JH, Noel R, Becker T, Cassidy LE, Clark S, Hughes SD, Newgard CB. Transfection of rat insulinoma cells with GLUT-2 confers both low and high affinity glucose-stimulated insulin release: relationship to glucokinase activity. J Biol Chem 1994; 269:11523—11529.

12. Miyazaki J-I, Araki K, Yamato E, Ikegami H, Asano T, Shibasaki Y, Oka Y, Yamamura K-I. Establishment of a pancreatic β cell line that retains glucose-inducible insulin secretion: special reference to expression of glucose transporter isoforms. Endocrinology 1990;127:126—132.

13. Asfari M, Janjic D, Meda P, Li G, Halban PA, Wollheim CB. Establishment of 2-mercaptoethanol-dependent differentiated insulin-secreting cell lines. Endocrinology 1992;130:167—178.

14. Efrat S, Leiser M, Surana M, Tal M, Fusco-Demane D, Fleischer N. Murine insulinoma cell line with normal glucose-regulated insulin secretion. Diabetes 1993;42:901—907.

15. Wang S, Beattie GM, Mally MI, Lopez AD, Hayek A, Levine F. Analysis of a human fetal pancreatic islet cell line. Transplantation Proc 1997;29:2219.

16. Clark SA, Quaade C, Constandy H, Hansen P, Halban P, Ferber S, Newgard CB, Normington K. Novel insulinoma cell lines produced by iterative engineering of GLUT-2, glucokinase, and human insulin expression. Diabetes 1997;46:958—967.

17. Hohmeier HE, BeltrandelRio H, Clark SA, Henkel-Reiger R, Normington K, Newgard CB. Regulation of insulin secretion from novel engineered insulinoma cell lines. Diabetes 1997;46: 968—977.

18. Mandrup-Poulsen T. The role of interleukin-1 in the pathogenesis of IDDM. Diabetologia 1996; 39:1005—1029.

19. Rabinovitch A. Roles of cytokines in IDDM pathogenesis and islet β-cell destruction. Diabetes Rev 1993;1:215—240.

20. Corbett JA, McDaniel M. Does nitric oxide mediate autoimmune destruction of β-cells? Possible therapeutic interventions in IDDM. Diabetes 1992;41:897—903.

21. Eizirik DL, Flodstrom M, Karlsen AE, Welsh N. The harmony of the spheres: inducible nitric oxide synthase and related genes in pancreatic beta cells. Diabetologia 1996;39:875—890.

22. Hohmeier HE, Thigpen A, Tran V, Davis R, Newgard CB. Stable expression of manganese superoxide dismutase (MnSOD) in insulinoma cells prevents IL-1β-induced cytotoxicity and reduces nitric oxide production. J Clin Invest 1998;101:1811—1820.

Bone-inducing implants: new synthetic absorbable poly-D,L-lactic acid-polyethylene glycol block copolymers as BMP-carriers

Kunio Takaoka[1], Naoto Saito[1], Shimpei Miyamoto[2], Hideki Yoshikawa[2] and Takao Okada[3]

[1]Department of Orthopaedic Surgery, Shinshu University School of Medicine, Nagano; [2]Department of Orthopaedic Surgery, Osaka University Medical School, Osaka; and [3]Central Institute for Research, Taki Chemical Co. Ltd., Hyogo, Japan

Abstract. Bone morphogenetic proteins (BMPs) are a group of proteins which are thought to be responsible for new bone formation seen in the process of fracture healing. Successful production of BMPs through recombinant DNA technology provides an opportunity to constitute new artificial implants with bone-inducing capacity by combining with synthetic implants. To build these bioactive implants, degradable polymer materials were synthesized, mixed with BMP and screened for their capacity to induce bone in an in vivo system. Implants consisting of polylactic acid (PLA) polymers and its derivative polymers with liquid and plastic properties in combination with rBMP consistently induced new bone in mice within a few weeks after implantation. Our data and the work of others suggest that these reconstituted bone-inducing materials have the potential to substitute bone grafts in the healing of fractures or repair of bone defects. The combination of this hybrid material with porous solid biomaterials could provide a bone-inducing implant with mechanical strength. Further preclinical and clinical work to check the safety of these implants will be necessary before they are adopted for widespread clinical use.

Keywords: bone morphogenetic protein, carrier, synthetic degradable polymer.

Introduction

In an attempt to develop new synthetic bone graft substitutes that can promote local bone formation, one possible approach is to combine growth factors with synthetic biomaterials [1]. Recent advances in genetic engineering have made it possible to produce many growth factors (proteins) which have potential clinical use. To fully realize the clinical value of these growth factors, safety checks and optimal delivery systems are essential for any development program. The purpose of this study is to develop a delivery system for a bone-inducing protein, bone morphogenetic protein (BMP), involving synthetic polymers which are compatible with host tissues and biodegradable, and do not modify the biological activity of the protein.

Address for correspondence: Dr Kunio Takaoka MD, Department of Orthopaedic Surgery, Shinshu University School of Medicine, Asahi 3-1-1, Matsumoto, Nagano 390-8621, Japan. Tel.: +81-263-37-2659. Fax: +81-263-35-8844.

BMPs

BMPs are a group of proteins which have the unique biological capacity to elicit new bone formation in vivo. The biological activity of BMP was originally found in decalcified bone matrix by an experimental model of ectopic bone formation [2]. In this experimental model, bone from an animal was decalcified in a 0.6 N HCl solution and a fragment of the decalcified bone was implanted into the muscle of an allogeneic animal. At 4—6 weeks after implantation, the decalcified bone matrix was resorbed and replaced by bone of host origin (Fig. 1). The biologically active factor, which is located in normal bone matrix and is responsible for this bone induction, is BMP [3]. BMP is also thought to be responsible for new bone formation seen in the process of fracture healing. In bone, the BMP molecules are produced by osteoblasts and retained in the bone matrix. BMP is also produced by transformed osteoblasts (osteosarcoma cells) and retained in osteosarcoma tissue [4]. We have purified BMP from a murine osteosarcoma using biochemical procedures [5]. The molecular size of the purified BMP was 32 kDa and it exists as a dimeric structure. Based on partial amino acid sequence information, the complementary DNA encoding the BMP molecule was cloned [6]. From the nucleotide sequence information of the c-DNA, the whole amino acid sequence of the pro-BMP molecule was determined and named BMP-4. Figure 2 shows the amino acid sequences of the human BMP-4

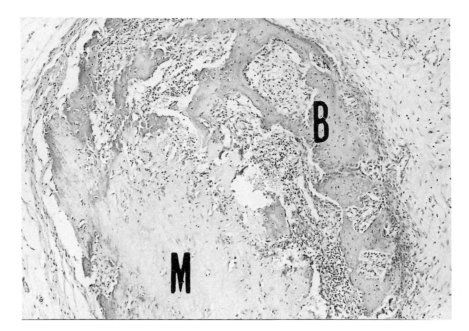

Fig. 1. Bone induction by decalcified bone matrix. HCl-decalcified bone matrix of a rat was implanted into the muscle of allogeneic rat. The matrix (M) was partially resorbed and replaced by bone (B) of host origin. The factor responsible for this bone induction is BMP.

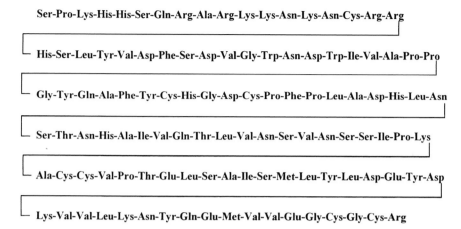

Fig. 2. Amino-acid sequence of BMP-4 human-type molecules.

monomers. Other BMP molecules except BMP-1 are known to have similar structures. A common feature of these BMP molecules is the position of the cysteine residues at the carboxyl terminus. The positions of these seven cysteine residues are the same as those for TGF-β, indicating that BMP molecules are members of the TGF-β superfamily [7]. The BMP molecules with biological activity have a dimeric structure; the monomer which has no biological activity has a molecular size of around 16 kDa and is processed to a biologically active homodimeric structure of 32 kDa through the creation of an intermolecular disulfide bond.

To date, some BMP molecules have been successfully synthesized by DNA recombination methodologies with the resultant protein products (rBMP) retaining the original biological bone-inducing activity. Large quantities of human BMPs (BMP-2, BMP-4, BMP-7, and GDF-5) are becoming available for clinical use. Figure 3 shows the bone mass induced ectopically in mice by BMP-2. rBMP-2 (5 µg) was combined with pure type-1 collagen (2 mg) and implanted into the muscles of a mouse. Three weeks after implantation, the implant was resorbed and replaced by bone.

Constitution of bone-inducing composite implant

Successful large-scale production of rBMP provides us with the possibility of constituting an implant with bone-forming capacity for medical use as an alternative to native bone graft materials. However, some problems have to be solved before rBMP is ready for use in clinical practice. One major hurdle is the development of an optimal delivery system for BMP to work effectively at the implanted site. In order to construct the delivery system, a suitable carrier material is essential. In previous studies, only when BMP was combined with a suitable carrier material, such as collagen, and implanted into muscles was new

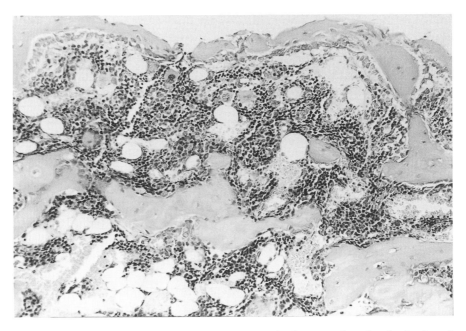

Fig. 3. Histology of a locally induced ossicle in the muscle of a mouse 3 weeks after implantation of the rBMP-2/collagen composite.

bone formed consistently [8,9]. These experimental data clearly highlight the essential role of a carrier for successful expression of the bone-inducing capacity of BMP.

Another important function of the carrier material is the regulation of the shape and size of the BMP-induced bone mass. When sufficient quantities of BMP are combined with a carrier molded to specific dimensions, a bone mass with a configuration resembling the original implant is formed [9]. Again, these results point to the essential role of the carrier as a determinant of size and shape of the induced bone mass. As already shown, collagen from animal sources can be used as a carrier material of BMP and has been utilized routinely in animal experiments and recently in clinical trials [9–23]. However, some potential risks or disadvantages associated with the use of xenogeneic animal collagen should be taken into consideration. The potential risks are disease transfer from animals as seen in spongy encephalitis and unexpected immune or inflammatory reactions from hosts, which, in turn, compromise BMP action [24–27]. To avoid these problems, we have focused on the development of novel synthetic polymers as BMP carriers.

Requirements of a BMP carrier

In general, the carrier materials for BMP are required to have a couple of special physicochemical and biological properties. The materials should not be inflammatory or cytotoxic because an intense inflammatory reaction impedes bone formation by BMP [28]. The carrier material should be insoluble under physiological conditions so as to retain the BMP molecules in situ, but susceptible to absorption by host tissue and replacement by the newly formed BMP-induced bone [29]. Synthetic carriers for BMP should meet these requirements [30–45].

Synthetic absorbable polymers as BMP-carriers

PLA polymers

As a candidate material with degradable properties, synthetic D,L,polylactic-acid homopolymers (PLA) were tested [42]. The physical properties, degradation rates and biological reactions from host tissues change depending on the molecular sizes of the PLA polymers. Therefore, PLA polymers with various molecular sizes were synthesized, mixed with BMP-4 and implanted in mice to test the new bone forming activity of the PLA/BMP composites. New bone formation was seen only at 3 weeks after implantation when PLA with a molecular size of 650 Da was implanted with BMP. In high molecular sized PLA/BMP composite implants, many foreign body giant cells and inflammatory cells were seen but no bone was formed. In the lowest molecular sized PLA (160 Da)/BMP composite implant, tissue necrosis and infiltration of inflammatory cells were seen again in the absence of bone formation. From these results, we concluded that PLA with a molecular size of 650 Da and a waxy nature at room temperature appeared to work as a carrier material for BMP. However, this PLA was thought less than optimal because the induced bone mass was significantly smaller than the original size of the implant. The reasons for the reduced efficacy of the PLA650 polymer as a BMP carrier are thought to be due to its rapid degradation and high acidic nature.

PLA-polyethylene glycol block copolymers

In order to improve the efficacy of the PLA650 polymer, a hydrophilic polyethylene glycol polymer (PEG) with a molecular size of 200 or 600 Da was linked to the PLA polymer (PLA-PEG block copolymer) (Fig. 4) by a condensation reaction [41]. The goals of this modification were to retard the degradation rate through an increase in molecular size and reduction in the acidity of the PLA650 polymer. By this modification, the degradation rate was retarded and the low pH value of PLA650 was much improved. By use of this modified polymer, performance of bone induction by BMP was improved (Fig. 5) and the induced bone mass with a constant dose of BMP became almost equivalent to the size of the original implant. This PLA-PEG block copolymer has a viscous liquid nature at room temperature and can be used as an injectable carrier material, but is difficult to mold to specific three-dimensional configurations.

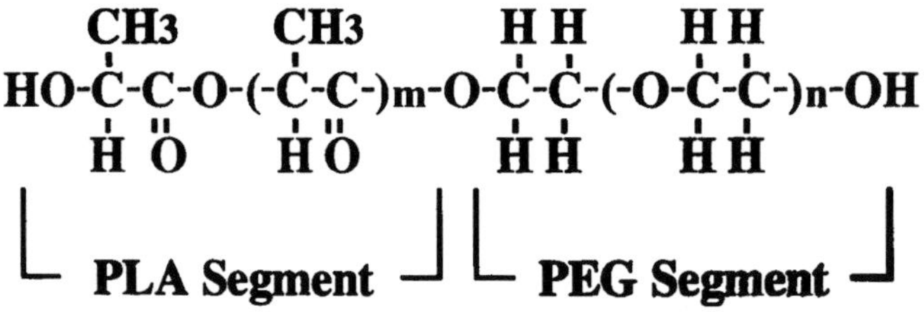

146

$$\text{HO-}\underset{\underset{\text{H}}{|}}{\overset{\overset{\text{CH3}}{|}}{\text{C}}}\text{-}\underset{\underset{\text{O}}{||}}{\text{C}}\text{-O-(-}\underset{\underset{\text{H}}{|}}{\overset{\overset{\text{CH3}}{|}}{\text{C}}}\text{-}\underset{\underset{\text{O}}{||}}{\text{C}}\text{-)m-O-}\underset{\underset{\text{H}}{|}}{\overset{\overset{\text{H}}{|}}{\text{C}}}\text{-}\underset{\underset{\text{H}}{|}}{\overset{\overset{\text{H}}{|}}{\text{C}}}\text{-(-O-}\underset{\underset{\text{H}}{|}}{\overset{\overset{\text{H}}{|}}{\text{C}}}\text{-}\underset{\underset{\text{H}}{|}}{\overset{\overset{\text{H}}{|}}{\text{C}}}\text{-)n-OH}$$

└─ **PLA Segment** ─┘ └─ **PEG Segment** ─┘

Fig. 4. Assumed chemical formula of PLA-PEG block polymers suitable for a carrier of BMP.

Plastic PLA-PEG-PLA block copolymer

Because of the difficulty in molding and handling the viscous liquid PLA650-PEG200 polymer, we developed new types of PLA-PEG-PLA block copolymers with a plastic nature which enables easy manipulation to mold into arbitrary configurations in intraoperative situations (Fig. 6). The basic chemical formula of this plastic carrier is shown in Fig. 7. The feature of this type of copolymer is that the PLA polymers are linked to both sides of the PEG polymer. To derive the plastic nature, the molecular sizes of both PLA segments and PEG segments were increased. The degradation rates for these types of copolymers tend to be retarded when compared with PLA650-PEG200 polymers.

In order to screen for the optimal molecular sizes of PLA and PEG blocks of this plastic type of copolymer, many types of PLA-PEG-PLA block copolymers with various molecular lengths and various PLA/PEG ratios were constructed and tested. These plastic block copolymers have a hydrophilic nature and swell when placed in water. Because of the plastic nature of this carrier, creating a

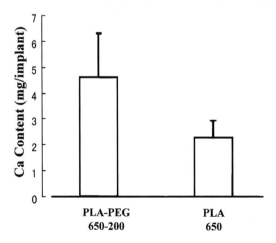

Fig. 5. Calcium contents of ossicles induced with BMP-4/PLA-650 or its derivative polymer (PLA650-PEG200 block copolymer). Process of BMP-induced bone formation was improved with use of the PLA-PEG polymer.

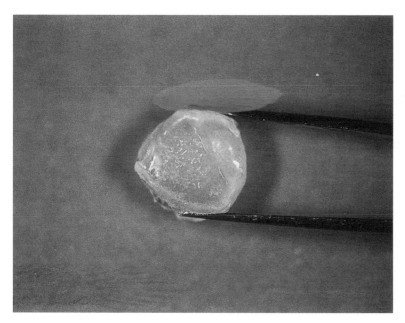

Fig. 6. Appearance of a soft and plastic PLA-PEG-PLA block polymer.

homogeneous mixture with BMP became difficult and a novel mixing procedure had to be developed. The procedure to combine BMP with these plastic polymers was as follows: each polymer was placed in a glass test tube and chilled acetone was added to dissolve the polymer; rBMP-2 dissolved in 0.01 N HCl was then added and mixed well with the dissolved polymer. The mixture was then centrifuged under vacuum to remove acetone and to restore the plastic nature of the carrier polymer. BMP/polymer composite implants prepared in this manner were implanted into the back muscles of mice and examined for bone-inducing activity. Three weeks after implantation, the implants were harvested and examined for new bone formation by radiological and histological methods. New bone formation was seen with great frequency when some of the PLA-PEG-PLA

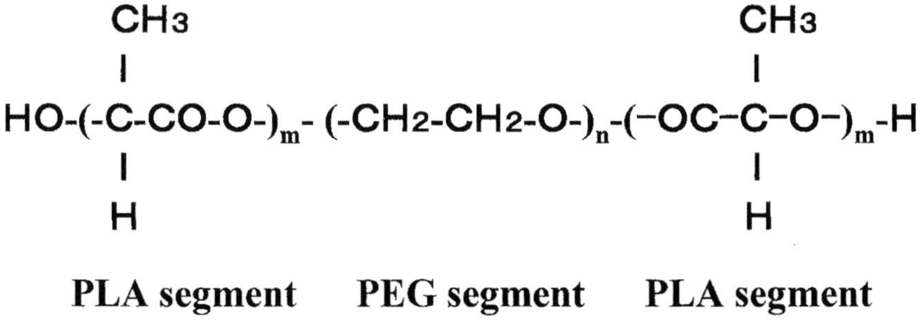

$$\text{HO-}(\text{-}\underset{\underset{\text{H}}{|}}{\overset{\overset{\text{CH}_3}{|}}{\text{C}}}\text{-CO-O-})_m\text{-} (\text{-CH}_2\text{-CH}_2\text{-O-})_n\text{-}(\text{-OC-}\underset{\underset{\text{H}}{|}}{\overset{\overset{\text{CH}_3}{|}}{\text{C}}}\text{-O-})_m\text{-H}$$

PLA segment   PEG segment   PLA segment

Fig. 7. Assumed chemical formula of plastic and degradable PLA-PEG-PLA block polymers.

block copolymers were used as the BMP carrier. From these experimental results, we concluded that the polymer composed of a single 3-kDa PEG segment and two 3.2-kDa PLA segments with a plastic nature worked well as a synthetic carrier of BMP and may have potential clinical use.

The expanding nature of this polymer in water might be beneficial in considering implantation into bone defects. By expansion after implantation, contact of the implant to bone will become tight and the dead space between the host bone and implant will be filled more easily after surgery. Another advantage of this expansion property of the implant will be in combination with surface porous biomaterials. When pores of the solid implant are filled with the BMP/polymer composite and implanted, the BMP/polymer composite will exude from the pores and form a layer covering the surface of the biomaterials. As a result, a layer of bone covering the biomaterials might be formed by BMP, which may encase the implant and enhance biological fixation or osseointegration of the biomaterials to the host bone. Figure 8 is an example of bone induction on the surface of a block of porous hydroxyapatite combined with BMP-2 and the plastic carrier polymer.

In order to substantiate the efficacy of rBMP/synthetic polymer composite in promoting bone healing in clinical use, further tests in large animals or primates will be essential before clinical trials of these bone-inducing implants can proceed.

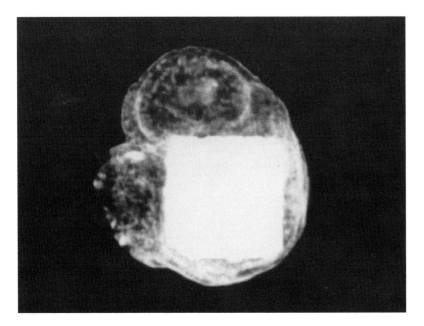

Fig. 8. BMP/polymer composite was packed into pores of a porous hydroxyapatite (HA) Block. At 3 weeks after the implantation of this combined implant into dorsal muscles of mouse, new ectopic bone completely surrounded the HA block.

References

1. Einhorn TS. Enhancement of fracture-healing. J Bone Joint Surg 1995;77(A):940—956.
2. Urist MR. Bone formation by autoinduction. Science 1965;150:893—899.
3. Urist MR, Iwata H. Preservation and biodegradation of the morphogenetic property of bone matrix. J Theor Biol 1973;38:155—161.
4. Amitani K, Nakata Y, Stevens J. Bone induction by lyophilized osteosarcoma in mice. Calcif Tis Res 1974;16:305—313.
5. Takaoka K, Yoshikawa H, Hashimoto J, Miyamoto S, Masuhara K, Nakahara H, Matsui M, Ono K. Purification and characterization of a bone-inducing protein from a murine osteosarcoma (Dunn type). Clin Orthop 1993;292:122—129.
6. Takaoka K, Yoshikawa H, Hashimoto J, Masuhara K, Miyamoto S, Suzuki S, Ono K, Matsui M, Oikawa S, Tsuruoka N, Tawaragi Y, Inuzuka C, Katayama T, Sugiyama M, Tsujimoto M, Nakanishi T, Nakazatio H. Gene cloning and expression of a bone morphogenetic protein derived from a murine osteosarcoma. Clin Orthop 1994;294:344—352.
7. Celeste AJ et al. Identification of transforming growth factor β family members present in bone inductive protein purified from bovine bone. Proc Natl Acad Sci USA 1990;87:9843—9847.
8. Takaoka K, Nakahara H, Yoshikawa H et al. Ectopic bone induction on and in porous hydroxy-apatite combined with collagen and bone morphogenetic protein. Clin Orthop 1988;234: 250—254.
9. Takaoka K, Koezuka M, Nakahara H. Telopeptide-depleted bovine skin collagen as a carrier for bone morphogenetic protein. J Orthop Res 1991;9:902—907.
10. Aspenberg P, Turek T. BMP-2 for intramuscular bone induction: effect in squirrel monkeys is dependent on implantation site. Acta Orthop Scand 1996;67:3—6.
11. Boden SD, Moskovitz PA, Morone MA, Toribitake Y. Video-assisted lateral intertransverse process arthrodesis: validation of a new minimally invasive lumber spinal fusion technique in the rabbit and non-human primate (rhesus) models. Spine 1996;21:1689—1697.
12. Schimandle JH, Boden SD, Hutton WC. Experimental spinal fusion with recombinant human bone morphogenetic protein-2. Spine 1995;20:1326—1337.
13. Rutherford RB, Spangberg L, Tucker M, Rueger D, Charette M. The time course of the induction of reparative dentine formation in monkeys by recombinant human osteogenic protein-1. Arch Oral Biol 1994;39:833—838.
14. Rutherford RB, Wahle J, Tucker M, Rueger D, Charette M. Induction of reparative dentine formation in monkey by recombinant human osteogenic protein-1. Arch Oral Biol 1993;38: 571—576.
15. Ripamonti U, Heliotis M, Rueger DC, Sampath TK. Induction of cementogenesis by recombinant human osteogenic protein-1 (hOP-1/BMP-7) in the baboon (Papio ursinus). Arch Oral Biol 1996;41:121—126.
16. Niederwanger M, Urist MR. Demineralized bone matrix supplied by bone banks for a carrier of recombinant human bone morphogenetic protein (rhBMP-2): a substitute for autogeneic bone grafts. J Oral Implant 1996;22:210—215.
17. Kirker-Head CA, Gerhart TN, Schelling SH, Hennig GE, Wang EW, Holtrop ME. Long-term healing of bone using recombinant human bone morphogenetic protein 2. Clin Orthop 1995; 318:222—230.
18. Cook SD, Wolfe MW, Salkeld SL, Rueger DC. Effect of recombinant human osteogenic protein-1 on healing of segmental defects in non-human primates. J Bone Joint Surg 1995;77(A): 734—750.
19. Cook SD, Baffes GC, Wolfe MW, Sampath TK, Rueger DC. Recombinant human bone morphogenetic protein-7 induces healing in a canine long-bone segmental defect model. Clin Orthop 1994;301:302—312.
20. Cook SD, Baffes GC, Wolfe MW, Sampath TK, Rueger DC, Whitecloud TS. The effect of recombinant human osteogenic protein-1 on healing of large segmental bone defects. J Bone Joint

Surg 1994;76(A):827—838.

21. Gerhart TN, Kirker-Head CA, Kriz MJ, Holtrop ME, Hennig GE, Hipp J, Schelling SH, Wang E. Healing segmental femoral defects in sheep using recombinant human bone morphogenetic protein. Clin Orthop 1993;293:317—326.

22. Yasko AW, Lane JM, Fellinger EJ, Rosen V, Wozeney JM, Wang EA. The healing of segmental bone defects, induced by recombinant human bone morphogenetic protein (rhBMP-2). J Bone Joint Surg 1992;74(A):659—670.

23. Toriumi DM, Kotlar HS, Luxenberg DP, Holtrop ME, Wang EA. Mandibular reconstruction with a recombinant bone inducing factor. Arch Otolaryngol Head Neck Surg 1991;117: 1101—1112.

24. Bach FH, Fishman JA, Daniels N, Proimos J, Anderson B, Carpenter CB, Forrow L, Robson SC, Finberg HV. Uncertainty in xenotransplantation: individual benefit versus collective risk. Nature Med 1998;4:141—144.

25. Back FH, Winkler H, Ferran C, Hancock WW, Robson SC. Delayed xenograft rejection. Immunol Today 1996;17:379—384.

26. Delustro F, Dasch JD, Keefe J, Ellingsworth L. Immune responses to allogeneic and xenogeneic implants of collagen and collagen derivatives. Clin Orthop 1990;260:263—279.

27. Furthmay H, Timple R. Immunochemistry of collagens and procollagens. Int Rev Connect Tis Res 1976;7:61—99.

28. Sela J, Applebaum J, Uretzky G. Osteogenesis induced by bone matrix is inhibited by inflammation. Biomat Med Dev Art Org 1986;14:227—237.

29. Nilsson OS, Persson P, Ekelund A. Heterotopic new bone formation causes resorption of the inductive bone matrix. Clin Orthop 1990;257:280—285.

30. Fischgrund JS, James SB, Chabot MC, Hankin R, Herkowitz HN, Wozney JM, Shirkhoda A. Augmentation of using rhBMP-2 and different carrier media in the canine spine fusion model. J Spinal Disord 1997;10:467—472.

31. Yamazaki Y, Oida S, Ishihara K, Nakabayashi N. Ectopic induction of cartilage and bone by bovine bone morphogenetic protein using a biodegradable polymeric reservoir. J Biomed Mater Res 1996;30:1—4.

32. Isobe M, Yamazaki Y, Oida S, Ishihara K, Nakabayashi N, Amagasa T. Bone morphogenetic protein encapsulated with a biodegradable and biocompatible polymer. J Biomed Mater Res 1996;32:433—438.

33. Mayer M, Hollinger J, Ron E, Wozeney J. Maxillary alveolar cleft repair in dogs using recombinant human bone morphogenetic protein-2 and polymer carrier. Plast Reconstr Surg 1996;98: 247—259.

34. Bostrom M, Lane JM, Tomin E, Browne M, Berberian W, Turek T, Smith J, Wozeney J, Schildhauer T. Use of bone morphogenetic protein-2 in the rabbit ulnar nonunion model. Clin Orthop 1996;327:272—282.

35. Agrawal CM, Best J, Hackman JD, Boyan BD. Protein release kinetics of a biodegradable implant for fracture nonunions. Biomaterials 1995;16:1255—1260.

36. Sandhu HS, Kanim LEA, Kabo JM, Toth JM, Zeegen EN, Liu D, Seeger LL, Dawson EG. Evaluation of rhBMP-2 with an OPLA carrier in a canine posterolateral (transverse process) spinal fusion model. Spine 1995;20:2669—2682.

37. Kenley R, Marden L, Turek T, Jin L, Ron E, Hollinger JO. Osseous regeneration in the rat calvarium using novel delivery system for recombinant human bone morphogenetic protein-2 (rhBMP-2). J Biomed Mater Res 1994;28:1139—1147.

38. Lee SC, Shea M, Battle MA, Kozitza K, Ron E, Turek T, Schaub RG, Hayes WC. Healing of large segmental defects in rat femurs is aided by rhBMP-2 in PLGA matrix. J Biomed Mater Res 1994;28:1149—1156.

39. Meikle MC, Papaioannou S, Ratledge TJ, Speight PM, Watt-Smith SR, Hill PA, Reynolds JJ. Effect of poly-DL-lactide-co-glycolide implants and xenogeneic bone matrix-derived growth factors on calvarial bone repair in the rabbit. Biomaterials 1994;15:513—521.

40. Miki T, Harada K, Imai Y, Enomoto S. Effect of freeze-dried poly-L-lactic acid discs mixed with bone morphogenetic protein on the healing of rat skull defects. J Oral Maxillofac Surg 1994; 52:387–391.

41. Miyamoto S, Takaoka K, Okada T, Yoshikawa H, Hashimoto J, Suzuki S, Ono K. Polylactic acid-polyethylene glycol block copolymer: a new biodegradable synthetic carrier for bone morphogenetic protein. Clin Orthop 1993;294:333–343.

42. Miyamoto S, Takaoka K, Okada T, Yoshikawa H, Hashimoto J, Suzuki S, Ono K et al. Evaluation of polylactic acid homopolymers as carriers for bone morphogenetic protein. Clin Orthop 1992;278:274–285.

43. Heckman JD, Boyan BD, Aufdemorte TB, Abbot JT. The use of bone morphogenetic protein in the treatment of nonunion in a canine model. J Bone Joint Surg 1991;73(A):750–764.

44. Lovell TP, Dawson EG, Nilsson OS, Urist MR. Augmentation of spinal fusion with bone morphogenetic protein in dogs. Clin Orthop 1989;243:266–274.

45. Kawamura M, Urist MR. Induction of callus formation by implants of bone morphogenetic protein and associated bone matrix noncollagenous proteins. Clin Orthop 1988;236:240–248.

Tissue Engineering for Therapeutic Use 3.
Y. Ikada and T. Okano, editors.

153

Tissue engineering in bone

Masaki Noda, Kunikazu Tsuji, Teruhito Yamashita, Nanako Kawaguchi,
Yoichi Ezura, Jinghong Li, Shunichi Murakami, Ichiro Sekiya, Yoshinori
Asou, Yuji Takazawa, Koichi Furuya, Ying Liu and Akira Nifuji
Department of Molecular Pharmacology, Medical Research Institute, Tokyo Medical and Dental University, Tokyo, Japan

Keywords: bone formation, bone resorption, osteoblasts, regeneration, transcription factor.

Introduction

Bone functions as a supporting tissue for the body of humans and animals, however, it also functions as a buffering system for the calcium metabolism which is tightly maintained in the body. Structural reconstruction of injured bones is a major issue in orthopedic surgery as well as in maxillofacial surgery. Bone grafting has been used for many practical cases. However, in the case of massive bone loss in certain parts of bones, it is necessary to contemplate bone regeneration in addition to the standard technique of bone grafting. Bone morphogenetic protein (BMP) was discovered in the 1950s and the availability of recombinant BMP has opened a new field for bone reconstruction and regeneration. Although the application of BMP is closer to clinical use than before, there are still problems to be solved with regard to the carriers for BMP. This issue has been under vigorous investigation. This review refers to the current understanding of the biological aspects of molecules recently identified as being involved in bone metabolism. These molecules are expected to contribute to find a novel pathway in contemplating tissue regeneration in bone. The review also describes the recently identified transcription factors related to skeletogenesis.

BMP and CBFA1

CBFA is one of the runt-domain-type transcription factors which were first discovered as a factor binding to polyoma virus enhancer (PEBP2α) [1−3]. Expression of this factor is detectable in differentiated F9 cells but not in undifferentiated cells. Alpha subunits consist of three isoforms. These heterodimerize with CBFB

Address for correspondence: Masaki Noda MD, PhD, Department of Molecular Pharmacology, Medical Research Institute, Tokyo Medical and Dental University, 3-10 Kanda-surugadai, 2-chome, Chiyoda-ku, Tokyo 101-0062, Japan. Tel./Fax: +81-3-5280-8066. E-mail: waka.mph@mri.tmd.ac.jp

(=PEBP2β) and bind to a consensus sequence comprising PuACCPuCA. Subsequently, CBFA1 transcription factor was found to be essential for the differentiation of osteoblasts since the knockout mice lacking the gene for CBFA1 do not develop bone at all [4,5]. These mice die within a few minutes after birth due to insufficiency of respiration. Histological investigation of these mice indicated that although thin layers of alkaline phosphatase positive cells were found in the bone, there was no accumulation of calcified bone matrix in the regions where bone should normally develop. Instead of bone development, cartilage formation was observed in most of the skeletons where endochondral bone formation takes place. However, bones which usually develop through a membranous bone formation pathway are totally lacking. In the heterozygous knockout mice, it was found that these mice lack a clavicle and at the same time it was found that human individuals who are known as patients with cleidocranial dysplasia possess mutations at multiple sites on the gene encoding CBFA1 [6,7]. In parallel to these observations, analyses of the promoter regions of osteoblastic-specific marker genes, encoding osteocalcin and osteopontin, identified specific response elements to osteoblast-specific factor, OSF2, which was later found to be identical to CBFA1.

Since BMP was known to be a potent promotion factor of osteoblastic differentiation, regulation of CBFA1 was tested in osteoblast cultures in the presence and absence of BMPs [8]. BMP was found to enhance the expression of CBFA1 gene in most of the cell culture systems indicating that CBFA is indeed downstream of BMP actions. As it was found before, the heterodimeric form of BMP, BMP4/7 is more potent than BMP2 in many biological assays including ectopic bone formation and *Xenopus* mesoderm induction. CBFA1 gene expression is also, more potently enhanced by the treatment with the heterodimeric form of BMP, BMP4/7 [9], compared to BMP2, indicating that similar hierarchy of BMP potency can be observed in the regulation of CBFA1. However, CBFA1 by itself is not the only downstream factor for BMPs as it was found that even by using the CBFA1−/− cells, BMP at high doses can induce the expression of osteoblastic marker genes such as osteocalcin and alkaline phosphatase [8]. Nonetheless, CBFA1 is still the necessary factor for osteoblastic differentiation.

We have investigated whether CBFA overexpression can modulate expression of downstream genes in osteoblastic and nonosteoblastic cells by using the expression vector for the shorter form of CBFA (CBFA1/PEBP2) [9]. This shorter form lacks about 80 amino acids on its N-terminal portion which is claimed to be specific to the form of the other isoform, CBFA1/OSF2. Overexpression of CBFA1/PEBP2 enhances the expression of osteopontin. However, at the same time it suppresses the expression of type I collagen gene both in osteoblastic cells and nonosteoblastic cells, indicating that the different isoforms of CBFA have different functions. These observations indicate that BMP acts through the expression of CBFA1. However, it would also act through alternative pathways, and use of BMP for bone regeneration should take these multiple pathways of osteoblastic differentiation into consideration to maximize the effect of the treatment with and application of growth factors.

Indian hedgehog as a possible growth factor involved in bone formation and its regulation

BMP has been known to play a significant role not only in bone formation but also in development of animals especially in the early period of embryonic morphogenesis [10–13]. One of the peptide growth factors, which is critical in morphogenesis of skeleton, is hedgehog. Indian hedgehog is one of the four hedgehog family members [14–19] and is found to be relatively specific in the skeletal tissue formation area. Indian hedgehog is expressed in prehypertrophic cartilage cells which are under the control of PTHrP (parathyroid-hormone-related peptide) [20]. Expression of Indian hedgehog is enhanced by PTHrP which in turn suppresses further differentiation of condrocytes.

We investigated whether Indian hedgehog is also expressed in the cells of osteoblastic lineage, and found that hedgehog is expressed, although at low levels, in osteoblast-like cells [21]. It was further shown that Indian hedgehog expression in osteoblastic cells is under the control of another potent modulator of osteoblastic function, transform growth factor β (TGFβ). TGFβ enhances the expression of Indian hedgehog gene in osteoblast-like cells [21]. This enhancement of Indian hedgehog gene expression by TGFβ is dependent on the elongation of the half-life of Indian hedgehog mRNA indicating post-translational control. We further investigated whether Indian hedgehog is also expressed in other types of osteoblastic cells and found that this gene is expressed in normal rat calvaria cells, and furthermore, TGFβ also enhances the expression of Indian hedgehog in the rat calvaria cells. These observations clearly suggest that Indian hedgehog is one of the target molecules which are under the control of TGFβ, a cytokine which is important for bone formation and regeneration.

BMP and skeletal morphogenesis

BMPs, including BMP2 and BMP4/7, are known to be expressed at the critical period of time and in the critical areas during early morphogenesis in embryos, such as the primordial tissues for the skeletons. Therefore, it is likely that BMP plays a role in determining the shape and size of bone, as well as the number of bones during the critical period of morphogenesis. Nifuji et al. investigated whether a minimal amount of BMP would be able to alter the shape of the skeleton by implanting carriers which can elicit BMP to the microenvironment where the cells sitting closely to the carrier would be affected. By these analyses, BMP was found to influence the shape of the embryonic skeletons [22]. The rib bones of chicken embryos have been observed to fuse or to make extra process or sometimes bifurcate in response to the implantation of BMP in the somites. Similarly, vertebral bones are also affected in their morphogenesis by the implantation of BMPs in somites at the early period of development [22]. In vertebrae, either the fusion of part of the process, or fusion with the rib was observed. Interestingly, there was a time window for the effect of BMPs in modulation of the morpho-

156

genesis of axial skeletons. When BMP was implanted on day 2 into the somites, it affected both vertebra and rib formation, however, only rib but not vertebra malformation was observed when implantation was conducted on day 3. These observations indicate that BMP could not only induce the differentiation of undifferentiated cells into the skeletal cells, but that it could also modulate the shape of the bones as a morphogenesis factor, suggesting that at a certain time point in the strategy of bone regeneration, we may also have to consider such activity of BMP.

BMP regulation of transcription factors

Since mesenchymal precursor cells differentiate into osteoblast, cartilage, muscle cells, and adipogenic cells, it was interesting to find out that an inhibitory transcription factor such as Id could be found in many types of cells including the ones derived from these mesenchymal stem cells. We previously reported that Id, which binds to and inhibits the action of helix-loop-helix (HLH)-type transcription factors, is expressed in osteoblastic cells and its regulation is under the control of many calciotropic factors [23–25]. Scleraxis is one of the HLH transcription factors which was expressed in sclerotome and appears to be involved in the formation of subsequent connective tissues, including skeletons. Investigations of the expression of scleraxis in skeletal cells indicated that this transcription factor is expressed in osteoblast-like cells and its expression is under the control of BMP [26]. Treatment with BMP of ROS72.8 cells suppresses the expression of scleraxis. Because scleraxis by itself can enhance the expression of aggrecan and also type II collagen, markers for chondogenic cells, it is somehow involved in the connective tissue cell differentiation which may not include direction of osteoblastic cells [27]. Suppression of scleraxis by BMP suggests that the BMP would redirect the pathway of the differentiation from the nonosteoblastic line to osteoblastic lines. We also found that TGFβ can enhance scleraxis expression in osteoblastic cells [28]. Furthermore, we found that FGF can also modulate scleraxis expression in skeletal cells [29]. These observations indicate that the HLH-type transcription factor, scleraxis, is the target of BMP and other osteotropic cytokines, and this molecule may determine the fate of the precursors for skeletal cells, although the exact nature of scleraxis function is still to be determined.

Summary

Tissue engineering has to be considered as a part of the effort to use host cells to reconstruct or to resume the function of the organs of targets including bone and cartilage. Since BMP is considered to be the closest molecule for clinical application, it is hoped that in the near future the problems with the carriers are solved and that the application of this molecule would be made possible. However, bone is also under the control of many other growth factors, and therefore, it is necessary to identify the interplay of the multiple factors to efficiently regenerate bone and cartilage in the body. Discovery and understanding of new molecules which could act as humoral or transcription factors would facilitate research in efficiently regenerating skeletal tissues in patients.

References

1. Bae SC, Yamaguchi-Iwai Y, Ogawa E, Maruyama M, Inuzuka M, Kagoshima H, Shigesada K, Satake M, Ito Y. Isolation of PEBP2 alpha B cDNA representing the mouse homolog of human acute myeloid leukemia gene, AML1. Oncogene 1993;8:809−814.
2. Ogawa E, Maruyama M, Kagoshima H, Inuzuka M, Lu J, Satake M, Shigesada K, Ito Y. PEBP2/PEAZ represents a family of transcription factors homologous to the products of the *Drosophila* runt gene and the human AML1 gene. Proc Natl Acad Sci USA 1993;90: 6859−6863.
3. Ogawa E, Inuzuka M, Maruyama M, Satake M, Naito-Fujimoto M, Ito Y, Shigesada K. Molecular cloning and characterization of PEBP2 beta, the heterodimeric partner of a novel *Drosophila* runt-related DNA binding protein PEBP2 alpha. Virology 1993;194:314−331.
4. Komori T, Yagi H, Nomura S, Yamaguchi A, Sasaki K, Deguchi K, Shimizu Y, Bronson RT, Gao YH, Inada M, Sato M, Okamoto R, Kitamura Y, Yoshiki S, Kishimoto T. Targeted disruption of Cbfa1 results in a complete lack of bone formation owing to maturational arrest of osteoblasts. Cell 1997;89:755−764.
5. Otto F, Thornell AP, Crompton T, Denzel A, Gilmour KC, Rosewell IR, Stamp GW, Beddington RS, Mundlos S, Olsen BR, Selby PB, Owen MJ. Cbfa1, a candidate gene for cleidocranial dysplasia syndrome, is essential for osteoblast differentiation and bone development. Cell 1997; 89:765−771.
6. Mundlos S, Otto F, Mundlos C, Mulliken JB, Aylsworth AS, Albright S, Lindhout D, Cole WG, Henn W, Knoll JH, Owen MJ, Mertelsmann R, Zabel BU, Olsen BR. Mutations involving the transcription factor CBFA1 cause cleidocranial dysplasia. Cell 1997;89:773−779.
7. Lee B, Thirunavukkarasu K, Zhou L, Pastore L, Baldini A, Hecht J, Geoffroy V, Ducy P, Karsenty G. Missense mutations abolishing DNA binding of the osteoblast-specific transcription factor OSF2/CBFA1 in cleidocranial dysplasia. Nature Genet 1997;16:307−310.
8. Ducy P, Zhang R, Geoffroy V, Ridall AL, Karsenty G. Osf2/Cbfa1: a transcriptional activator of osteoblast differentiation. Cell 1997;89:747−754.
9. Tsuji K, Ito Y, Noda M. Expression of the PEBP2aA/AML3/CBFA1 gene is regulated by BMP4/7 heterodimer and its overexpression suppresses type I collagen and osteocalcin gene expression in osteoblastic and nonosteoblastic mesenchymal cells. Bone 1998;22:87−92.
10. Ferguson EL, Anderson KV. Decapentaplegic acts as a morphogen to organize dorsal-ventral pattern in the *Drosophila* embryo. Cell 1992;71:451−461.
11. Padgett RW, Wozney JM, Gelbart WN. Human BMP sequences can confer normal dorsal-ventral patterning in the *Drosophila* embryo. Proc Natl Acad Sci USA 1993;90:2905−2909.
12. Re'em-Kalma Y, Lamb T, Frank D. Competition between noggin and bone morphogenetic protein 4 activities may regulate dorsalization during *Xenopus* development. Proc Natl Acad Sci USA 1995;92:12141−12145.
13. Leim KF, Tremml G, Rokelink H, Jessell TM. Dorsal differentiation of neural plate cells induced by BMP-mediated signals from epidermal ectoderm. Cell 1995;82:969−979.
14. Riddle RD, Johnson RL, Laufer E, Tabin C. Sonic hedgehog mediates the polarizing activity of the ZPA. Cell 1993;75:1401−1416.
15. Roelink H, Augsburger A, Heemskerk J, Korzh V, Norlin S, Ruiz I, Altaba A, Tanabe Y, Placzek M, Edlund T, Jessell TM, Dodd I. Floor plate and motor neuron induction by vhh-1, a vertebrate homolog of hedgehog expressed by the notochord. Cell 1994;76:761−775.
16. Chang DT, Lopez A, von-Kessler DP, Chiang C, Simandl BK, Zhao R, Seldin MF, Fallon JF, Beachy PA. Products, genetic linkage and limb patterning activity of a murine hedgehog gene.

Development 1994;120:3339—3353.

17. Johnson RL, Laufer E, Riddle RD, Tabin C. Ectopic expression of sonic hedgehog alters dorsal-ventral patterning of somites. Cell 1994;79:1165—1173.

18. Fan CM, Tessier-Lavigne M. Patterning of mammalian somites by surface ectoderm and noto-chord: evidence for sclerotome induction by a hedgehog homolog. Cell 1994;79:1175—1186.

19. Bitgood MJ, McMahon AP. Hedgehog and Bmp genes are coexpressed at many diverse sites of cell-cell interaction in the mouse embryo. Devel Biol Chem 1995;172:126—138.

20. Vortkamp A, Lee K, Lanske B, Segre GV, Kronenberg HM, Tabin CJ. Regulation of rate of cartilage differentiation by Indian hedgehog and PTH-related protein. Science 1996;273:613—622.

21. Murakami S, Nifuji A, Noda M. Expression of Indian hedgehog in osteoblasts and its post-transcriptional regulation by transforming growth factor-β. Endocrinology 1997;138:1972—1978.

22. Nifuji A, Kellermann O, Kuboki Y, Wozney JM, Noda M. Perturbation of BMP signaling in somitogenesis resulted in vertebral and rib malformations in the axial skeletal formation. J Bone Min Res 1997;12:332—342.

23. Ogata T, Noda M. Expression of Id, a member of HLH protein family, is downregulated at con-fluence and enhanced by dexamethasone in mouse osteoblastic cell line, MC3T3E1. Biochem Biophys Res Comm 1991;180:1194—1199.

24. Kawaguchi N, DeLuca HF, Noda M. Id gene expression and its suppression by 1,25-dihydroxy-vitamin D_3 in rat osteoblastic osteosarcoma cells. Proc Natl Acad Sci USA 1993;89:4569—4572.

25. Ogata T, Wozney J, Benezra R, DeLuca HF, Noda M. A differentiation promoting factor, bone morphogenetic protein-2 enhances expression of a helix-loop-helix molecule, Id (inhibitor of differentiation) in proliferating osteoblast-like cells. Proc Natl Acad Sci USA 1993;90:9219—9222.

26. Liu Y, Wozney J, Nifuji A, Olson E, Noda M. Scleraxis is expressed in C2C12 myoblasts and its level is downregulated by bone morphogenetic protein-2 (BMP2). J Cell Biochem 1997;67:66—74.

27. Liu Y, Watanabe H, Nifuji A, Olson E, Noda M. Overexpression of a single helix-loop-helix type transcription factor, scleraxis, enhances aggrecan gene expression in osteoblastic osteosarcoma ROS17/2.8 cells. J Biol Chem 1997;272:29880—29885.

28. Liu Y, Nifuji A, Olson E, Noda M. Sclerotome-related helix-loop-helix type transcription factor (scleraxis) mRNA is expressed in osteoblasts and its level is enhanced by type-beta transforming growth factor. J Endocrinol 1996;151:491—499.

29. Kawauchi T, Mataga N, Tamura M, Olson E, Bonaventure J, Shinomiya K, Liu Y, Nifuji A, Noda M. Fibroblast growth factor downregulates expression of a basic helix-loop-helix-type transcription factor, scleraxis, in a chondrocyte-like cell line, TC6. J Cell Biochem 1998;70:468—477.

Index of authors

Adachi, E. 109
Agbaria, R. 119
Akaike, T. 87
Ando, K. 1
Asou, Y. 153
Audet, J. 15

Clark, S. 133
Cohen, S. 119
Conneally, E. 15

Eaves, C. 15
Eaves, A. 15
Endo, K. 71
Ezura, Y. 153

Fukuda, J. 43
Funatsu, K. 43
Furuya, K. 153

Gion, T. 43
Glicklis, R. 119

Han, H.-P. 133
Hayashi, T. 109
Hiratsuka, S. 25
Hirose, M. 109
Hohmeier, H.E. 133
Honda, Y. 77
Hotta, T. 1

Ide, C. 77
Ijima, H. 43
Ikada, Y. 71
Imamura, Y. 109
Inokuchi, S. 1
Interewicz, K. 87

Kan, P. 53
Kaneko, M. 43
Karamuk, E. 87
Kato, S. 1
Kawada, H. 1
Kawaguchi, N. 153
Kikuchi, A. 99
Kim, S.W. 99

Kimura, M. 1
Kitahara, A.K. 71
Kiyotani, T. 71
Kohsaka, S. 99
Kosugi, H. 109
Kruse, F. 133

Lee, G. 71
Li, J. 153
Liu, Y. 153

Maru, Y. 25
Mayer, J. 87
Merchuk, J.C. 119
Miura, N. 35
Miyamoto, S. 141
Miyatake, H. 1
Miyoshi, H. 53
Mizoguchi, A. 77
Mizuno, K. 109
Murakami, S. 153

Nakamura, Y. 1
Nakamura, T. 71
Nakano, I. 77
Nakazato, K. 109
Nakazawa, K. 43
Newgard, C.B. 133
Nifuji, A. 153
Nishida, A. 77
Nishimura, Y. 71
Noda, M. 153
Noda, T. 25

Ogawa, S. 25
Ohnishi, K. 71
Ohshima, N. 53
Okada, T. 141
Okano, T. 99, 109

Piret, J. 15

Quaade, C. 133

Raisman, G. 65
Rose-John, S. 15

160

Saito, N. 141
Sakurai, Y. 99
Sawano, A. 25
Schuppin, G. 133
Sekiya, I. 153
Shapiro, L. 119
Shibuya, M. 25
Shimada, M. 43
Shimakura, Y. 1
Shimizu, Y. 71
Shimizu, T. 1
Shirabe, K. 43
Sugimachi, K. 43
Sugiyama, T. 35
Sumida, Y. 109
Suzuki, K. 71
Suzuki, Y. 71

Takahashi, T. 25
Takahashi, J. 77
Takahashi, M. 77
Takahashi, S. 109
Takaoka, K. 141
Takazawa, Y. 153
Takeda, Y. 109

Takenaka, K. 43
Takimoto, Y. 71
Teramachi, M. 71
Terasaki, T. 99
Thigpen, A. 133
Tomihata, K. 71
Tsuji, K. 153
Tsuji, T. 1

Vien Tran, V. 133
von Recum, H.A. 99

Wintermantel, E. 87

Yabana, N. 25
Yamaguchi, S. 25
Yamashita, T. 153
Yamato, M. 99, 109
Yanagi, K. 53
Yasui, O. 35
Yoshikawa, K. 109
Yoshikawa, H. 141

Zmora, S. 119

Keyword index

alginate sponges 119
angiogenesis 25
axon 65

biofunctionalization 87
blood progenitors 15
bone
  formation 153
  marrow transplantation 15
  morphogenetic protein 141
  resorption 153

carrier 141
cell
  culture 119
  differentiation 110
  lines 133
  scaffolds 119
  transplantation 35, 99
co-culture 53, 99
collagen fibrils type V 110
collagen gel
  type I 110
  type IV 110
collagen tube 71
composites 87
cytokines 1

degradables 87
differentiation 77
diffusion 87

endothelial cell 25
extracorporeal circulation 43

functional recovery 71

gelatin tube 71
gene
  therapy 35
  transfer 1
genetic engineering 133
growth factors 15

hematopoietic system 1

hepatocyte 35, 53, 119
  spheroid culture 43
hybrid artificial organ 99

insulin 133

LEC rat 35
long-distance gaps 71

molecular biology 133

neural stem cells 77
nonparenchymal liver cell 53

osteoblasts 153
oval cell 35

perfusion 87
poly(N-isopropylacrylamide) 99
polyglycolic acid (PGA)-collagen tube 71
polyurethane foam 43
polyvinyl formal resin 53
primary hepatocyte spheroid 43

regeneration 1, 65, 153
retina 77
retrovirus 1

scaffolds 87
sciatic nerve 71
self-renewal 15
shear stress 53
signal transduction 25
smooth muscle cell 110
spinal cord 65
stromal cell line 1
synthetic degradable polymer 141

temperature-responsive 99
tissue engineering 43
transcription factor 153
transplant 77
transplantation 133
tyrosine kinase receptor 25

VEGF 25